Designing for Strength

Principles and Practical Aspects of Stress
Analysis for Engineers and Students

Peter Polak

Department of Mechanical Engineering,
University of Sheffield

First published 1982 by
THE MACMILLAN PRESS LTD
London and Basingstoke
Companies and representatives
throughout the world

Typeset in 10/12 Times by
Oxprint Ltd, Oxford

Printed in Hong Kong

ISBN 0 333 32674 1
ISBN 0 333 32676 8 pbk

CONTENTS

PREFACE

This book is intended to cover strength calculations for general design purposes at a standard suitable for most three-year engineering degree courses, but omitting academic fracture mechanics, soil mechanics, reinforced concrete, advanced civil engineering, aircraft structures and plasticity. The basic plan is to explain the main stress forms first, leading on to a critical examination of material tests and properties. Then before proceeding to applications it is deemed necessary to present a substantial summary of stress concentration data. The applications aspect is treated first in fairly fundamental terms, followed by aspects and procedures specific to structural steelwork and pressure vessels. It concludes with case studies, further discussion of earlier items, some useful tables, derivations and graphical methods, etc. Within some of the chapters the natural break-points between elementary and advanced work have been indicated. Most readers will wish to take a quick look at the advanced sections but defer detailed study of the arguments until a second work-through.

A large amount of important recent material and little-known older material is included in suitably condensed form, with references to readily available publications, as may be seen from the list. These references have all been closely studied by the author. An unusual feature is the inclusion of buckling as a major stress form. Most books leave this to the end as if it were a minor academic refinement instead of a major design criterion. Designers should welcome the frequent reminders to consider the highest equivalent stress, out-of-plane shear and Poisson contraction.

Sections containing particularly recent or rarely seen material of practical relevance are 2.4.2, 3.4.3, 4.4, 4.5, 5.4, 6.4.2, and 7.6 (three supports), 7.8 to 7.10 particularly flange bending effects, 7.12, 8.3.4, 8.5, 9.3, 9.8 and 9.9.

In engineering science we often find that full-size work, analogue or small-scale experiments and calculations all give different results. Sometimes these can be tracked down to one of the following

(1) genuine size effect, notably the difficulty of scaling object size and material grain size in the same ratio without altering some properties;

(2) Poisson's ratio differences, notably in photoelastic three-dimensional work;

(3) loading differences between specimens in testing machines and details within real structures;

(4) communication problems – many experiments and their logging are carried out by assistants rather than by the authors of the papers; this leaves room for misunderstandings, oversights and insufficient precautions against systematic errors;

(5) computing problems – many papers give the very basic equations and the conclusions without describing the routes between them; in computation it is often necessary to simplify the route and it is not always obvious what physical features have been concealed thereby.

It is always difficult to steer a middle course between long tedious explanations and too much brevity. In case the book is used without the benefit of teaching staff, the earlier parts are explained fairly fully. In the advanced sections it is assumed that the reader has also advanced in knowledge and familiarity with the material, so the explanations are more condensed in the interests of smoothness. The basic examples are set out in some detail to help in avoiding misunderstandings; advanced examples are treated more concisely, as befits more advanced students.

Bearing in mind that this is a degree-standard text with much new material, it is hoped that the presentation adopted will please most of the people most of the time.

ACKNOWLEDGEMENTS

The author wishes to thank the University of Sheffield for the granting of one term's study leave, photographic and secretarial assistance and library services. The help of the library of the Institution of Mechanical Engineers is also gratefully acknowledged.

The following have kindly granted permission to reproduce material: the editors of *Engineering, Engineering Materials and Design* and *Aircraft Engineering*; also the following publishers: The Institution of Mechanical Engineers, Chapman & Hall, Plenum Press, The British Standards Institution, Verlag Stahleisen (Dusseldorf), The British Constructional Steelwork Association and CONSTRADO.

Special thanks are due to Alan Chidgey for a number of improvements in chapter 9, consisting of additional points, rectification of misleading impressions, etc.

NOTATION

A	cross-sectional area usually with subscript; axial length of flanges
a	general-purpose length notation
B	breadth
b	general-purpose length notation
C	speed of stress waves as specified locally; torsional stress constant (torque/τ); spring constant D/d
D, d	diameters
E	Young's modulus
E'	wide-beam modulus $E/(1-\nu^2)$
e	extension, eccentricity of loading
F	force
f	frequency (spring surge); f_f = flexibility factor; f_s = stress factor (pipe bends)
G	modulus of rigidity
g	gravitational acceleration
H	height of beam section
h	general height; sag of catenary
I	second moment of area
J	polar moment
K	torsional stiffness constant for non-circular sections
k	radius of gyration, usually k_y or the lowest (oblique in angle sections), $k = \sqrt{(I/A)}$, also stiffners or thermal conductivity
L	length, span of beam
M	bending moment; M_f = fixing moment; M_c = central moment; M_j = junction moment (corner moment)
m	coefficient used in flange theory
N	number of active turns in helical spring
n	end-fixing constant in Euler strut calculation
P	force, usually vertical point load; P_{ce} = critical elastic buckling load
p	fluid pressure; contact pressure; force per unit contact area
Q	force, usually horizontal
R	resilience, dx/dP; reaction force; radius of curvature or large radius
r	radius
S, s	distance along a curve
T	torque; tension in catenary; thickness (larger of two thicknesses)
t	thickness; temperature

U	strain energy
V	volume
w	load per unit length
x,y	cartesian coordinates, defined locally if necessary
Z	section modulus, I/y_{max}
z	defined locally as required
α	coefficient of linear thermal expansion mm/mm °C
Δ	deflection
ϵ,e	strain
θ	angle
ν	Poisson's ratio
ρ	density
σ	direct stress, usually tensile if positive (except when discussing buckling)
σ_y	yield stress except in sections 2.5, 2.6 and A.11 (defined locally as stress in y direction)
σ_{ce}	mean compressive stress when member is at critical elastic buckling load
σ_d	design stress (pressure vessels, chapter 9)
σ_r	radial stress
τ	shear stress; τ_{ce} = critical elastic stress for shear buckling of beam web
ϕ	shear angle, general angle if defined locally
ψ	angle
ω	angular velocity

SUMMARY OF USEFUL FORMULAE

Some are accurate, others approximate for quick-checking work.

I = second moment of area (moment of inertia) as used in bending equations

I/y, sometimes called Z = section modulus in bending (elastic) when $\sigma = M/Z$

J = polar moment as used in torsion formulae for circular shafts

J/r = torsional modulus convenient for finding stess when τ = torque $\times r/J$

K, C are torsional constants for long non-circular bars, equivalent to J and J/r

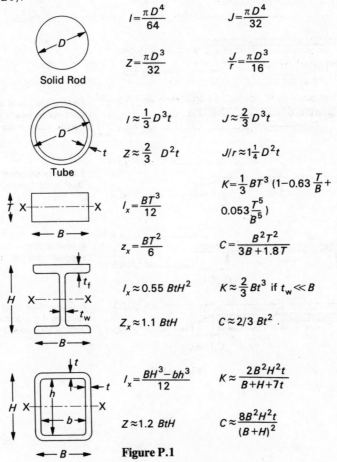

$$I = \frac{\pi D^4}{64} \qquad J = \frac{\pi D^4}{32}$$

$$Z = \frac{\pi D^3}{32} \qquad \frac{J}{r} = \frac{\pi D^3}{16}$$

Solid Rod

$$I \approx \frac{1}{3} D^3 t \qquad J \approx \frac{2}{3} D^3 t$$

$$Z \approx \frac{2}{3} D^2 t \qquad J/r \approx 1\frac{1}{4} D^2 t$$

Tube

$$I_x = \frac{BT^3}{12} \qquad K = \frac{1}{3} BT^3 \left(1 - 0.63\frac{T}{B} + 0.053\frac{T^5}{B^5}\right)$$

$$z_x = \frac{BT^2}{6} \qquad C = \frac{B^2 T^2}{3B + 1.8T}$$

$$I_x \approx 0.55\, BtH^2 \qquad K \approx \frac{2}{3} Bt^3 \text{ if } t_w \ll B$$

$$Z_x \approx 1.1\, BtH \qquad C \approx 2/3\, Bt^2$$

$$I_x = \frac{BH^3 - bh^3}{12} \qquad K \approx \frac{2B^2 H^2 t}{B + H + 7t}$$

$$Z \approx 1.2\, BtH \qquad C \approx \frac{8B^2 H^2 t}{(B + H)^2}$$

Figure P.1

SELECTED BENDING MOMENT FORMULAE FROM CHAPTER 7

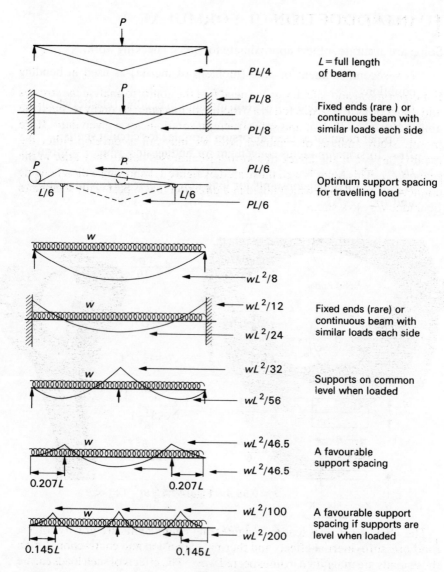

Figure P.2

1 INTRODUCTION

For economical and safe design we need first the ability to analyse the stresses and deflections to be expected in a structure or machine; secondly we need the ability to select efficient and safe structural solutions for a given duty. If we merely study analysis of standard cases we may fail to develop either the creative aspect or the knack of spotting weaknesses before they arise in the real object. This point is well illustrated by figure 1.1 showing a diesel engine connecting-rod design which failed in a direction that superficially seemed to be under very low stress.[1]

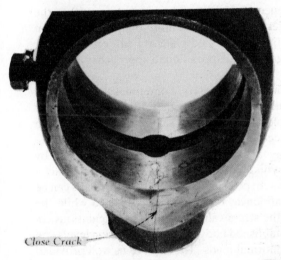

Close Crack

Figure 1.1

The chief sources of load on engineering components are gravity forces, fluid pressure, inertial effects and thermal expansion and contraction. Often these loads are magnified in unexpected ways. The effects of such loads can be classified as tension, compression (often with danger of buckling), bending, shearing and torsion. Figure 1.2 shows what we mean by these terms, using familiar examples.

For clarity's sake we start by treating these aspects separately, which implies some simplifying assumptions since they hardly ever occur alone. Although we shall produce some apparently accurate treatments, in the

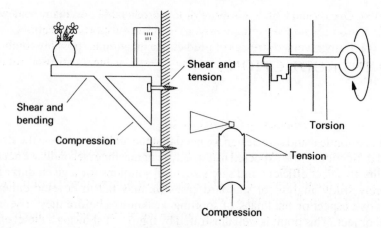

Figure 1.2

author's many years of experience in design work it is rare for either the loading or the material properties to be known to more than two significant figures. The dimensions of castings and rolled plate are subject to similar uncertainties; hence correct diagnosis and upper-bound approaches are more important than accurate theory.

Designs always include considerations of safety. Sometimes these take the form of a large numerical factor which is often merely a cloak for poor attempts at analysis. An example of a superficial safety factor of 12 which proved to be unsafe appears in section 10.2

In building structures the safety factor is included in the loading figures, by assuming a load in excess of any likely loading, in conjunction with only a small factor applied to the stresses. In pressure vessels, where the loads can be predicted with considerable confidence, realistic loading is used while the safety factor is incorporated in the stress values used in the calculations. In most industries there is a well-established body of customary stress levels and ways of assessing normal and abnormal loads. These serve us well provided that new designs are within the range of previous experience. When going beyond this, it may be helpful to remember the main headings that influence the choice of materials and working stresses.

(a) Load Conditions

(1) Maximum load known, repeated occasionally, as in pressure vessels in continuous processes, crane and bridge structures, steam power plant casings, etc.

(2) Maximum load known, repeated frequently, as in batch processing vessels, conveyor plant, aircraft at take-off, engines at full power, etc.

(3) Occasional high loads more or less predictable plus many somewhat lesser loads repeated very frequently, as in most transport applications.

(4) Fully reversed stresses of predictable magnitude, possibly combined with steady stresses, as in most classes of heavy or highly stressed rotating machinery.

(b) *Adverse Environmental Conditions*

(1) Sustained high temperature and/or corrosive environment, with very infrequent shutdowns.

(2) Frequent temperature changes and/or alternating exposure to two media, for example, air and water.

(3) Low temperature conditions involving brittle fracture danger.

(4) Low temperature conditions involving extra loads due to ice build-up.

(c) *Internal Forces*

(1) Thermal stresses due to welds not stress-relieved, heat-treatment, grinding, faulty machining, uneven temperatures in service especially on start-up (unfavourable).

(2) Residual stress due to yielding during a high first load (may be favourable in many cases).

(3) Residual compressive stresses due to surface rolling, peening, carburising, etc. (mostly favourable).

(d) *Service Conditions (in Decreasing Order of Severity)*

(1) Inaccessible to inspection; this may occur in dams, concrete-cased steel members in buildings, plant which becomes radioactive, main cables of suspension bridges.

(2) Inspected annually or every few years.

(3) Accessible to frequent inspection.

(4) Instrumented or otherwise continuously monitored, for example, machinery which runs roughly as soon as defects develop, pipes and cables subject to leak detection or electrical checks.

(e) *Failure Consequences (in Decreasing Order of Severity)*

(1) Major human disaster.

(2) Fatal accident or severe injury to public.

(3) Fatal accident or severe injury to operators.

(4) Major inconvenience with danger, for example, failure of power in

hospitals, ventilation failure in underground railway, water supply failure, etc.

(5) Serious ecological disturbance, for example, a large oil spill.

(6) Major inconvenience without direct danger, for example, commuter traffic jam, seaway blockage.

(7) Loss of reputation and goodwill due to persistent failures.

(8) Dissatisfied customer alleging consequential loss.

(9) Dissatisfied customer demanding money back.

(10) Downgrading to lower permitted loads/speeds.

The causes of failure are wrong assessment of the stresses, unforeseen misuse, faulty manufacture, faulty material, unsuitable material specification. Many failures seem to be due to multiple causes, often involving much consequential damage which obscures the causes.

2 STRESSES AND STRAINS

Although many objects are designed on the basis of rigidity with very low general stress levels, the designer is responsible for providing sufficient strength at all points while avoiding wasteful use of materials, often without any chance of building and testing a series of prototypes. For a rational design we need to relate loads, sizes and stresses by calculation, together with deciding on suitable working stress values. These are based on standardised tests and we must understand their uses and limitations. In this chapter we discuss the basic stress forms; this prepares the ground for a critical discussion of properties and the way data are obtained.

2.1 TENSION AND COMPRESSION

There are many engineering components subject to more or less pure tension, including bolts, ropes, electricity transmission lines, pressure vessel shells, chains and belts. In compression we have pillars, arches, valve push-rods, ball bearings, jacks.

It is obvious that the strength of a component in tension is proportional to the cross-sectional area, a large cross-section being equivalent to several smaller ones sharing the load. Thus we speak of a tensile stress, force/cross-sectional area normal to the load, called σ. Applying a tensile force to a component increases its length and slightly reduces its cross-section. The actual extension is proportional to the original length and in metals it is proportional to the load. It is perhaps unfortunate that the most familiar components whose extension is large enough to be noticeable are ropes and rubber bands, both of which have the property of extending readily at first and stiffening up at higher loads. This is due to their structure: as the fibres or molecules get more highly aligned they become stiffer and more like metals. Up to a certain value of stress, the extension is transitory and the object returns to its previous length upon unloading.

The extension per unit length is called the tensile strain, ϵ. A reversible or transitory strain is called elastic; any permanent extension comes under the heading of plastic flow. In the elastic region the ratio of stress to strain is Young's modulus, E

$$E = \frac{\sigma}{\epsilon} \tag{2.1}$$

If we have a member of length L, cross-sectional area A and apply a force P to it in tension, it will extend by an amount

$$\Delta = \frac{PL}{AE} \tag{2.2}$$

This in itself is not very useful since we rarely have a uniform component; however the equation can be applied to the various portions, adding up the extensions. In mathematical language

$$\Delta = \frac{\Sigma PL}{AE} \quad \text{or} \quad \Delta = \frac{\Sigma L\sigma}{E} \tag{2.3}$$

The variable E as in rubber, plastics or rope is not dealt with here.

In compression the elastic behaviour is, in short pieces, the opposite of tensile behaviour, permanent deformation starting at similar stresses. Thereafter the component gets thicker and therefore stands greater loads. However in many practical cases there is a substantial difference. While in tension a component is self-straightening; in compression any lack of straightness gives rise to a bending moment tending to aggravate the problem. This action is called buckling. Since it involves bending, discussion must be deferred for the moment. It is mentioned here because of its important design implications.

Example 2.1

The bolt shown in figure 2.1 is made of material with a yield point of 500 N/mm² and Young's modulus 2×10^5 N/mm². Neglecting the effect of fillet radii and stress concentrations, find the force P at which the bolt will start to yield and also the elastic extension between A and B while loaded with this force.

Smallest cross-section = $\frac{1}{4}\pi \times 12^2$ mm², hence yield will start at a load, P, of $500 \times \frac{1}{4}\pi \times 12^2 = 56\,500$ N. Then

$$\text{extension} = \frac{\Sigma PL}{AE} = \frac{4P}{\pi E} \times \left(\frac{28}{12.5^2} + \frac{32}{14^2} + \frac{50}{12^2} \right) = 0.25 \text{ mm}$$

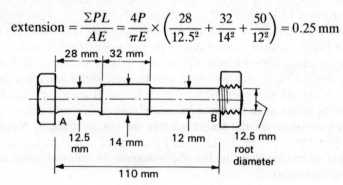

Figure 2.1

Problem 2.1

Figure 2.2 shows a steel bar used to support the walls of a building. Its diameter is 24 mm with threaded portions of 3 mm pitch, 19.8 mm root diameter. For elastic extension purposes the effective diameter of the threaded portion is more like 21 mm. How many turns of one nut are required to develop a force of 60 kN in the bar if the walls remain rigid? What is the highest stress in the bar, ignoring stress concentrations due to sharp corners etc? Take E as 200 000 N/mm².

Ans. 1.1 turns, 195 N/mm².

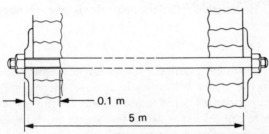

Figure 2.2

Problem 2.2

A motor vehicle valve push-rod exerts a force of 600 N. It is made of steel tube of 10 mm outside diameter, 0.4 mm wall thickness and is 250 mm long. By how much is it compressed? $E = 200\ 000$ N/mm².

Ans. 0.06 mm.

Problem 2.3

A concrete building is 60 m high. The weight is carried on columns running the full height of the structure, each column carrying a total load equal to 4 times its own mass. If the concrete's density is 2200 kg/m³ and its Young's modulus is 2×10^{10} N/m², show that the shortening due to gravity is of the order of 7.8 mm if the stress varies uniformly from top to bottom. Remember gravity.

2.2 BENDING

Bending is the most common and most varied form of loading. Usually we design to avoid plastic flow, therefore we consider mainly elastic cases; in bending there can be a significantly different situation when plastic flow develops, producing an additional safety margin. We shall discuss this at the end of the section.

2.2.1 The Elastic Bending Equation

To describe pure bending we need a pure couple, the bending moment; the beginner should find it easier to visualise a simpler situation such as a cantilever with a point load, corresponding to a diving-board or fishing-rod. Analytically speaking a cantilever is more complex than we would wish because it has a shearing force but we shall ignore this for the present.

Figure 2.3 shows a rubber cantilever, inscribed with orthogonal grid lines. We note that except near the fixing the grid lines still cross substantially at right angles. The top is extended, the bottom compressed. The layer which remains unchanged in length is called the neutral axis. Shear is negligible most of the way.

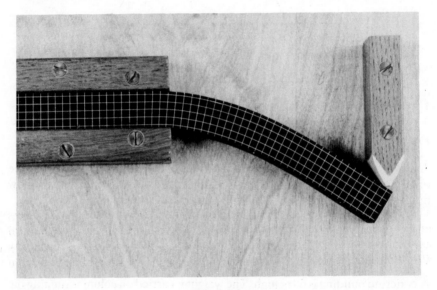

Figure 2.3

To analyse this situation, consider figure 2.4. The two planes AABB and CCDD were parallel to each other when the beam was straight, a distance x apart. Now that the beam is bent, these planes converge at OO. The radius of curvature is R. Taking a typical layer of thickness dy, at a distance y above the neutral axis n: the layer is of length $x + dx$, so it is under strain of dx/x and therefore under stress $E\,dx/x$ just like any member in tension.

By similar triangles, $dx/x = y/R$, hence

$$\text{stress } \sigma = \frac{Ey}{R} \tag{2.4}$$

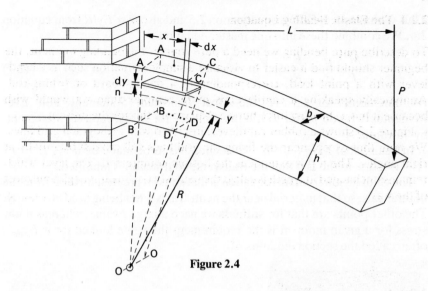

Figure 2.4

The layer exerts a force of stress × area = $(Ey/R) \times b\,dy$.

Now we take moments about the neutral axis

$$\text{moment exerted by the layer} = \text{force} \times y = \frac{E}{R}\,by^2\,dy$$

$$\text{moment due to the load} = PL$$

The opposing moment is the integral of all the forces exerted by the layers under tension above the neutral axis and also the layers below the neutral axis, in compression. The sign convention of positive tensile stress, positive y above the axis, means that we need not change signs, we simply integrate from $-h/2$ to $+h/2$

$$PL = \frac{E}{R}\int_{-h/2}^{h/2} by^2\,dy$$

But $\int_{-h/2}^{h/2} by^2\,dy$ is the second moment of area or moment of inertia of the cross-section, commonly called I, then

$$PL = \frac{EI}{R} \tag{2.5}$$

The bending moment need not be due to a single load P at a distance L but can be made up of many loads distributed over various distances; the total moment due to all loads beyond the point being considered is simply described as the bending moment, M, where

$$M = \Sigma PL \tag{2.6}$$

Thus we have $M = EI/R$ from equation 2.5 and also $\sigma = Ey/R$ from equation 2.4. We combine these into one master equation

$$\frac{M}{I} = \frac{\sigma}{y} = \frac{E}{R} \tag{2.7}$$

(A mnemonic for this is: the *M*ost *I*mportant *S*tatement *Y*ou *E*ver *R*emember! If you prefer to use the symbol f for stress, replace 'Statement' by 'Formula'.)

In beams that have $b \gg h$ a correction has to be made by using $E' = E/(1 - v^2)$, due to the fact that the transverse contraction is inhibited.

The chief lesson from this equation is that stress is greatest furthest from the neutral axis, so that material near the neutral axis is not being used efficiently. The other points are that for stiffness we need a large I value, whereas if low stress for a given moment is the requirement the value looked for is I/y_{max}, often called the section modulus, Z.

2.2.2 Various Beam Sections

The I value for a rectangular section as in figure 2.4 is $bh^3/12$. For a solid cylinder it is $\pi D^4/64$, where D is the diameter. Tubular sections have an I value obtained by subtracting the inner (absent) cylinder from the outer, so that if the outer diameter is D and the inner is d, then $I = \pi(D^4 - d^4)/64$. A similar argument applies to I-beams: if an I-beam is of width b, height h, wall thickness t, then $I = bh^3/12 - (b-t)(h-2t)^3/12$ which represents taking the whole occupied rectangle and subtracting from it the two rectangles taken away as shown in figure 2.5. A hollow rectangular beam would be treated in a similar way.

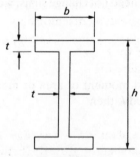

Figure 2.5

A comparison of sections is shown in figure 2.6. These are almost identical in I/y_{max} value so that for a given permissible stress value they would all be able to withstand the same bending moment. The I value then indicates their relative stiffness while the cross-sectional areas indicate economy of material. It also roughly indicates relative cost, with the proviso that making a more complex form requires somewhat more work and machinery.

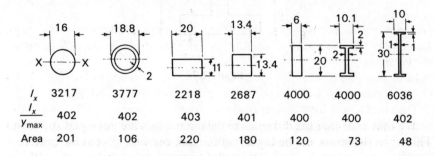

I_x	3217	3777	2218	2687	4000	4000	6036
$\dfrac{I_x}{y_{max}}$	402	402	403	401	400	400	402
Area	201	106	220	180	120	73	48

Figure 2.6

Sometimes we require the I value in the lateral direction; we distinguish between the main value as discussed above and the lateral value by calling the former I_x, the latter I_y. For the section shown in figure 2.5 the value I_y about the vertical axis of symmetry would be obtained by addition rather than subtraction as if we had separate rectangular bars: $I_y = 2tb^3/12 + (b - 2t)t^3/12$.

Unsymmetrical sections are treated on the same principles, but for convenience we can use the parallel axes theorem for second moments

I about any axis = I about the centroid
+ area × (distance from chosen axis to centroid)^2

This is illustrated by using a tee-section (figure 2.7). First we find the distance from the lower edge to the centroid; by first moments

$$\text{area} \times \bar{y} = (20 - 3.5) \times 4 \times 2 + 3.5 \times 30 \times 15$$

$$\text{area} = 66 + 105, \text{ therefore}$$

$$\bar{y} = \frac{1707}{66 + 105} = 9.98$$

Then we take second moments about the lower edge, I_E. For a rectangle of width b, height h, the second moment = $\int_0^h by^2 \, dy$ (between limits 0 and h) = $bh^3/3$

$$I_E = (20 - 3.5) \times 4^3/3 + 3.5 \times 30^3/3 = 31\,852$$

Then

$$I_x = I_E - \text{area} \times (\bar{y})^2 = 31\,852 - 171 \times 9.98^2 = 14\,820$$

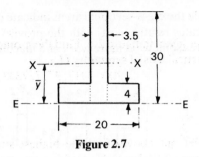

Figure 2.7

We may note that the distances to the outer edges are not equal above and below, so the stress will be highest at the top; the wide part at the bottom is under-used except in cases where the material has different strengths in tension and compression.

Examples like this are bound to be very simple. When we have discussed the remaining basic stress situations we shall return to bending, giving attention to the many types of structural forms and loading cases found in engineering.

Example 2.2

Find the highest bending stress, in N/mm^2, in a board 50 mm thick, 300 mm wide, carrying a load of 55 kg at a point 3 m beyond the fixing

$$\text{Bending moment} = 55 \times 9.81 \times 3000 \text{ N mm}$$

$$\sigma = \frac{My_{max}}{I} = \frac{55 \times 9.81 \times 3000 \times 25}{300 \times 50^3/12}$$

$$= \frac{55 \times 9.81 \times 3000}{300 \times 50^2/6} = 12.9 \text{ N/mm}^2$$

Example 2.3

A steel tape measure 0.08 mm thick has a curved profile of radius 20 mm when extended. When coiled up its smallest diameter is 40 mm but the profile is forced into the flat state. If $E = 2 \times 10^5 \text{ N/mm}^2$, find the highest stresses in the coiled state, ignoring interactive effects.

Coiling to a 20 mm radius is equivalent to flattening from a 20 mm radius, hence the stresses will be the same from both causes but acting in different directions. $\sigma/y_{max} = E/R$

$$\sigma = \frac{2.10^5 \times 0.08}{2 \times 20} = 400 \text{ N/mm}^2$$

Problem 2.4

Confirm all the values given in figure 2.6. Find the moment of inertia of these profiles about the vertical axis of symmetry (I_y).

Ans. 3217, 3777, 7333, 2687, 360, 354, 169.

Problem 2.5

Using the method set out above, find the highest stress in a member of cross-section as in figure 2.8 when subjected to a bending moment of 5×10^4 N mm.

Ans. 55.9 N/mm²; $\bar{y} = 6.73$ mm.

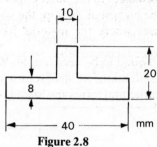

Figure 2.8

Problem 2.6

Find the highest stress under a bending moment of 600 N m in a round bar of 40 mm diameter and in a tube of 66 mm outer diameter and 52.5 mm inner diameter. Confirm that these contain the same amount of material per unit length.

Ans. 95.5, 35.45 N/mm².

2.2.3 Bending with Tension, Offset Tension/Compression

Sections 2.2.3, 2.2.4 and 2.2.5 may be omitted at a first reading.

Bending can occur in combination with direct force, usually due to direct loads offset from the centroid of the section. Up to the yield point the stresses are additive. As a simple example we take a member of tee-section as in figure 2.7 and apply a tension to it at the mid-point (*not* the centroid). Interpreting the dimensions as millimetres, let the force be 1000 N (figure 2.9a). This is equivalent to a force at the centroid plus a couple due to the offset. We note that $I = 14\ 820$ mm⁴, $\bar{y} = 9.98$ mm

$$\text{The direct stress} = \frac{1000}{A} = \frac{1000}{66 + 105} = 5.85 \text{ N/mm}^2$$

The bending stresses are My/I where y is either 9.98 mm or $30 - 9.98$ mm; $M = P \times e$

The tensile bending stress, at the top, =

$$\frac{1000 \times (15 - 9.98) \times 20.02}{14820} = 6.78 \text{ N/mm}^2$$

The compressive stress at the bottom =

$$\frac{1000 \times (15 - 9.98) \times 9.98}{14820} = 3.38 \text{ N/mm}^2$$

Adding the direct tension shows that all the stresses are tensile, ranging from 12.6 N/mm² at the top to 2.5 N/mm² at the bottom.

It is interesting to note that if we cut off the flange sides leaving only the central portion 30 mm high, 3.5 mm wide, the highest stress would be reduced. Wherever the force is compressive, we should look out for buckling of the thin free edges (see chapter 8).

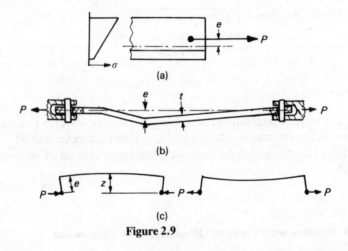

(a)

(b)

(c)

Figure 2.9

Another instructive example is a strut or tie with a small kink (see also section 7.13). We need to estimate the offset between the centre-line of the member and the load line; this we call e. The member is a rectangular bar, thickness t, width b normal to the paper. The direct stress $= P/bt$; the bending stress $My/I = M/Z = 6\,Pe/bt^2$ (figure 2.9b).

The greatest stress is the sum of these; expressed in terms of the direct stress

$$\sigma_{max} = \left(\frac{1 + 6e}{t}\right) \frac{P}{bt}$$

This clearly shows the importance of keeping members fairly straight.

If the force is compressive we must consider whether the elastic deflection due to the force augments the bending moment appreciably. The amount of bend will depend on how much of the length is subject to the bending moment. The worst case is when the whole length is off-centre, giving the eccentric strut, figure 2.9c.

The proof is tedious; we require the greatest deflection z, which of course gives the greatest bending moment, in terms of the initial eccentricity e

$$z = e \sec \left[\tfrac{1}{2}\sqrt{\left(\frac{PL^2}{EI}\right)} \right]$$

The stresses are found by using this value to give the greatest bending stress and then superposing the direct stress. If the member is in tension, $z = e \operatorname{sech} [\tfrac{1}{2}\sqrt{(PL^2/EI)}]$, the highest bending moment is then at the corner, of magnitude Pe.

2.2.4 Plastic Bending (Second year material)

If we apply so much bending moment that the surface stress exceeds the elastic limit, the simple analysis fails. For simplicity we assume that above some given stress value σ_y there is a sudden change, extension taking place at constant stress. This is essentially what happens in materials with a clear yield point. A comparison between the limiting elastic case, a typical case of yielding and the extreme case is shown in figures 2.10 a, b and c respectively. Case c cannot actually happen because we cannot produce large extensions right up to the neutral axis. However, assuming case c to happen, we take moments about the neutral plane by analogy with figure 2.4.

Force in slice $= \sigma_y b \, dy$

Moment due to this force $= \sigma_y by \, dy$

integrating, with due attention to sign reversal

Total moment $= \int \sigma_y by \, dy = 2\sigma_y \int_0^{t/2} by \, dy$ (2.8)

This equation is quite general, not restricted to the rectangular section shown. The mathematically minded will recognise the integral as the first moment of area for each half, above and below. The true first moment of area is of course zero about the centroid. The integral as used above, $2\int_0^{y_{max}} by \, dy$, is called the plastic modulus, for symmetrical sections. For a rectangle, this is easily evaluated as

$$2[\tfrac{1}{2} by^2]_0^{t/2} = \frac{bt^2}{4}$$ (2.9)

For an I-beam section it can be worked out by taking the overall rectangle value and subtracting the value for the rectangles taken away, or alternatively by dividing the section into suitable areas and taking each such area as acting at its centroid.

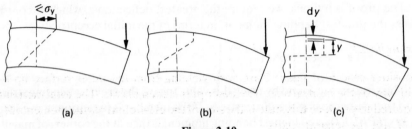

Figure 2.10

The practical use of the plastic modulus lies in assessing the deflections of structures when loaded beyond the point of initial yielding; in such cases several points in a structure can form so-called plastic hinges, possibly preventing total collapse.

Example 2.4

Find the plastic modulus of a circular section of diameter D.
   In figure 2.11, $b = 2[(D/2)^2 - y^2]^{1/2}$, by Pythagoras' theorem

$$\text{Plastic modulus} = 2\int by\,dy, \text{ limits } 0, \frac{D}{2} = \frac{4}{3}\left[\left(\frac{D}{2}\right)^2 - y^2\right]^{1\frac{1}{2}}$$

$$\text{Putting in the limits, plastic modulus} = \frac{D^3}{6} \tag{2.10}$$

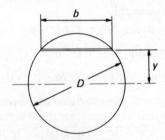

Figure 2.11

Problem 2.7

Find the plastic moduli about the x axis of the I-beam sections in figure 2.6.
                                                                    Ans. 492, 486.

2.2.5 Semi-plastic Bending

Since the fully plastic case is unrealistic, let us see briefly how to assess a mixed state. It is necessary to decide how far the yielding may be allowed to penetrate.

Example 2.5

Consider case b of figure 2.10, with yielding from the outer surface up to mid-way to the neutral axis. The inner part is then elastic. The total moment required to produce this state is the sum of the elastic and plastic moments M_e + M_p for the separate parts

$$M_e = \frac{\sigma_y \times b(t/2)^3}{(12 \times t/4)} = \frac{\sigma_y bt^2}{24} \qquad \begin{array}{l}(= \text{moment to produce } \sigma_y \\ \text{at boundary})\end{array}$$

$$M_p = \sigma_y \left[\frac{bt^2}{4} - \frac{b(t/2)^2}{4} \right] = \frac{\sigma_y \times 3bt^2}{16}$$

$$M_{\text{TOT}} = \frac{11\sigma_y bt^2}{48}$$

Example 2.6

Consider the last I-beam in figure 2.6 with the web elastic and the flanges yielding, that is, the yield boundary coinciding with the junction of flanges and web. The flanges are at uniform stress, acting at their centroids, the web acts as a simple elastic rectangle just reaching yield stress at the outer edges. Also find the full plastic moment.

$$M_e = \frac{\sigma_y \times 1 \times 28^3}{(12 \times 14)} = 131 \, \sigma_y$$

$$M_p = \sigma_y \times 2 \times 10 \times 14.5 = 290 \, \sigma_y$$

$$M_{\text{tot}} = 421 \, \sigma_y \qquad \text{Semi-plastic moment} = 421 \text{ units}$$

Full plastic moment for flanges is 290 units as above and for web = $bt^2/4 = 1 \times 28^2/4 = 196$. Total plastic moment = $196 + 290 = 486$ and ratio plastic/elastic = 1.21.

It should not surprise us that the plastic moment of an efficient section is little more than the elastic moment; for many rolled I-beam sections the ratio is only 1.1 or thereabouts, since most of the material is in the high-stress region anyway.

Problem 2.8

Find the moment for plastic flange and elastic web for the penultimate I-beam section in figure 2.6 (20 high, 10.1 wide, 2 thick), also the full plastic moment and the ratio full plastic/elastic.

Ans. 449, 492 units, 1.23.

2.3 BUCKLING

In common language, a buckled component implies any major distortion. In a design context we use the word buckling for a particular and very important set of cases of compressive loading, namely all those instances where a member in compression deflects so much that the bending moments due to the loss of straightness are sufficient to cause serious further deflections. The stresses may be elastic or may extend to the plastic region.

2.3.1 Elastic Buckling

We start with a case which is relatively rare in practice but easy to demonstrate and analyse; the more practical instances are then easily built up from this basic case. Consider a rod, initially straight, compressed by a force P until it bows out as shown in figure 2.12, the basic Euler strut. At any point, the bending moment acting on the rod is $M = Ph$. From the bending formula we have $M/I = E/R$. For simplicity we say $\mathrm{d}^2h/\mathrm{d}x^2 = -1/R$. This may be derived thus

$$\text{Slope} = \tan \theta$$

$$\frac{\mathrm{d}^2h}{\mathrm{d}x^2} = \frac{\mathrm{d}}{\mathrm{d}x(\text{slope})} = \sec^2 \theta \frac{\mathrm{d}\theta}{\mathrm{d}x}$$

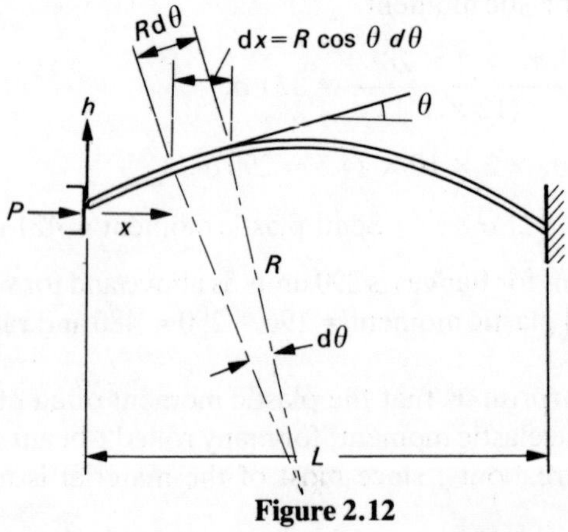

Figure 2.12

From the diagram, $dx = R \cos \theta\, d\theta$ so

$$\frac{d^2h}{dx^2} = \frac{-\sec^3 \theta}{R}$$

The negative sign is because the slope is diminishing. The error involved in making $\sec^3 \theta = 1$ is less than 6 per cent up to 0.2 radians.

Thus at any point

$$\frac{Ph}{I} = -E\frac{d^2h}{dx^2} \qquad\qquad (2.11)$$

The equation $d^2h/dx^2 = -(P/EI)h$ has the solution $h = A \sin \omega x + B \cos \omega x$, where $\omega = \sqrt{(P/EI)}$. By inspection

$$h = 0 \text{ at } x = 0 \text{ therefore } B = 0$$

$$h = 0 \text{ at } x = L, \text{ hence } \sin \omega L = 0$$

$$\omega L = \pi, \qquad \omega = \frac{\pi}{L}$$

Thus

$$h = A \sin \omega x = A \sin\frac{\pi x}{L}$$

$$\frac{d^2h}{dx^2} = -\left(\frac{\pi}{L}\right)^2 A \sin\frac{\pi x}{L} = -\left(\frac{\pi}{L}\right)^2 h$$

Substituting this in equation 2.11 gives $Ph/I = -Eh(\pi/L)^2$

$$P = \frac{\pi^2 EI}{L^2} \qquad \text{regardless of } h \qquad\qquad (2.12)$$

Our notation for the critical elastic force is P_{ce}.

The physical meaning behind this solution, seeing that it is independent of h, is that if we apply a load P greater than $\pi^2 EI/L^2$, the deflection starts to increase with no further increase of force. This is shown more clearly in figure 2.13; a plastic strip with rounded ends just started to deflect at the calculated load of 1¼ lb; when the load was increased to 1½ lb the strip was almost ready to bow freely. The next increase of load took the deflection to the full distance available on the apparatus.

This classic case is rare in practice; in structures we would not make compression members deliberately hinged at the ends except in certain cases of subsidence-resisting or earthquake-resisting structures. A striking instance of failure of a pin-jointed strut was reported in reference 2. Figure 2.14 shows how some very long, telescopic hydraulic jacks were used to re-position prefabricated ship sections. The movement starts with the left jacks short, the right jacks extended. The right jacks are allowed to close gradually as the

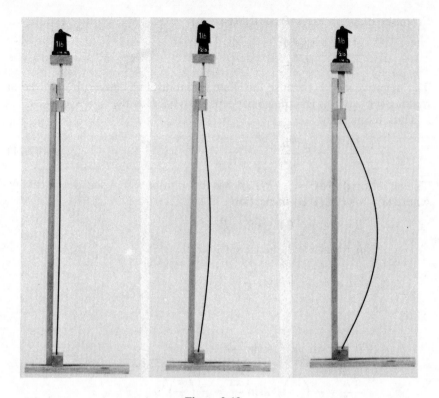

Figure 2.13

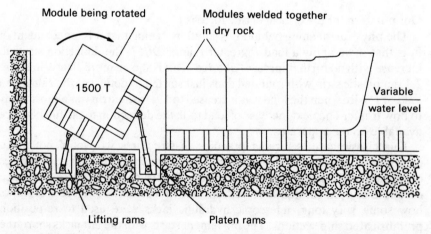

Figure 2.14 Reproduced by permission of *Engineering Materials and Design*, London

weight is gradually transferred to them. As P gets larger, L gets less so that $\pi^2 EI/L^2$ is always comfortably larger than the actual P; the critical or crippling condition is avoided. What happened one day was that the hydraulic oil was contaminated by some water and this water froze, blocking the outflow so that the right jacks remained extended; then as the weight was transferred to them, they buckled. In telescopic jacks the buckling tendency is aggravated by the unavoidable clearance in the sliding joints which allows some out-of-straightness to appear in addition to the bending of the components.

Figure 2.15 shows crippling forces for various other end conditions from

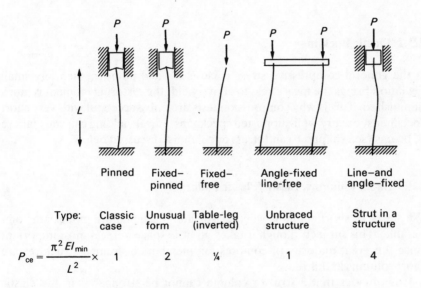

Some critical loads in buckling — keep well away

Figure 2.15

reference 3. They arise from the same basic sinusoidal equation. Figure 2.16 shows how the last two cases may manifest themselves in a structure. These are the commonest cases in engineering apart from the pinned–pinned basic case occurring in hydraulic jacks. In detail the hydraulic jack is more like two table-leg cases back to back, one being the ram, the other the body. The failing load would be lower than either of the components alone.

It is helpful to think of the buckling force P_{ce} (critical elastic) as not a force but a resistance to failure, since it is a property of the member, not of the loading system. It is customary to design members so that the resistance to elastic buckling is 3 times the expected loading. This allows for uncertainty of initial straightness to some extent.

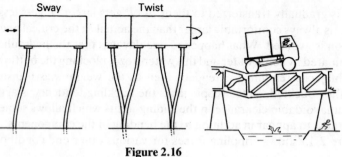

Figure 2.16

2.3.2 Plastic Buckling

If the general compressive stress is close to yield point σ_y then any small deviation brings the local stress to above yield; the restoring moment is much diminished. This is why compression tests are only successful with very short specimens, or very well-supported specimens. Local yielding can start failure of beams (see chapter 8) and struts or columns, discussed below.

2.3.3 Short Columns, Elastic/Plastic Failure

We speak of short columns when elastic and plastic failure modes are comparable. This subject, though it takes up little space here, is most important since the great majority of compression members in framed structures are short columns in this sense.

It is obvious that a strut or column cannot be stronger than the elastic crippling force P_{ce}. Equally obviously it cannot be stronger than $A\sigma_y$ since if we apply a load greater than $A\sigma_y$ the member simply starts squashing. These modes interact; if yielding is taking place, the resistance to bending vanishes. If yielding is imminent, then incipient bending will start it off and this will allow more bending. We take the Rankine approach: this amounts to saying that yield and buckling are parallel routes to disaster so we treat them like parallel resistances, adding their reciprocals.

$$\text{The probable failing force, } P_f = \frac{1}{1/P_{ce} + 1/A\sigma_y} \qquad (2.13a)$$

This formula is more convenient if transposed into terms of stress. Taking the critical buckling force calculated elastically as $n\pi^2 EI/L^2$, where n is the constant for various end fixings, namely ¼, 1, 2 or 4 as in figure 2.15, then since $I = Ak^2$ where k is the smallest radius of gyration, the elastic crippling stress $\sigma_{ce} = P_{ce}/A$ and the yielding stress σ_y. The failing stress σ_f then is

$$\sigma_f = \frac{1}{1/\sigma_y + 1/\sigma_{ce}} \tag{2.13b}$$

or $$\sigma_f = \frac{1}{1/\sigma_y + L^2/n\pi^2 Ek^2} \tag{2.13c}$$

There is a further form of buckling, in thin-walled fabricated work. This is wavy buckling of the flat parts, discussed in section 8.2.2. Rolled steel sections are designed with sufficiently thick parts to ensure that some yielding occurs before this form of buckling would appear. It can take place in plate-girder-type columns built up from sheets and sections, in very strong aluminium alloys where the low E and high σ_y values come together. By itself it is not a collapse mode but indirectly it contributes to collapse because the waviness leaves the members much less effective in compression than they should be, thus raising the maximum compressive stress for a given load.

Example 2.7

The BCSA structural steelwork handbook states the safe loads for various columns, for example, for a column profile of 327×311 mm area $= 201.2$ cm², $k_x = 13.9$ cm, $k_y = 7.89$ cm, flange 25 mm thick, web 15.7 mm. When rolled in steel of $\sigma_y = 425$ N/mm² the stated safe load on a 10 m high column is 1220 kN. Let us compare this with the calculated failing load under various conditions. Note the mixed units used in metricated structural tables!

For steel we can take $E = 200\ 000$ N/mm².

(1) Assume a properly braced structure, $n = 4$; $\sigma_{ce} = 4\pi^2 E\,k^2_y/L^2 = 492$ N/mm²

$$\sigma_f = \frac{1}{(1/492 + 1/425)} = 228\ \text{N/mm}^2$$

The wavy mode is negligible when the flange only extends by 6 times its thickness from the centre

$$P_f = 228 \times 201.2 \times 100\ \text{N} = 4587\ \text{kN}$$

(2) Assume no bracing; $n = 1$, $\sigma_{ce} = 492/4 = 123$ N/mm²

$$\sigma_f = \frac{1}{(1/123 + 1/425)} = 95.4\ \text{N/mm}^2 \qquad P_f = 1919\ \text{kN}$$

(3) Assume totally ineffective top fastening, that is, during erection or repair: fixed–free case, $n = \frac{1}{4}$, $\sigma_{ce} = 30.75$ N/mm²

$$\sigma_f = 28.67\ \text{N/mm}^2 \qquad P_f = 577\ \text{kN, less than half the safe load}$$

No wonder that erection and repairs tend to produce accidents.

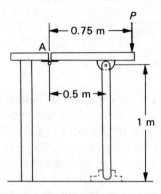

Figure 2.17

Example 2.8

Figure 2.17 shows a board hinged at A and supported by a round wooden rod of 20 mm diameter. If E for the wood is 10 000 N/mm², find the load P which is likely to cause collapse (a) if the leg is rounded at the base, (b) if it is fitted with a firm extended flat portion. Use elastic analysis only in this example.

The crippling load for condition a is $\pi^2 EI/L^2$

$$I = \frac{\pi d^4}{64}$$

$$P_c = \frac{\pi^2 \times 10\ 000 \times \pi \times 20^4}{64 \times 1000^2} = 775 \text{ N}$$

$$P = \frac{P_c \times 0.5}{0.75} \qquad \text{(by moments about A)}$$

Therefore

$$P = 517 \text{ N}$$

For condition b the crippling load is twice the above (from figure 2.15) hence $P = 1034$ N.

Example 2.9

Figure 2.18 shows a desk-top supported at each end by a strut of hard brass 12 mm deep, 7 mm thick, 400 mm long. If the strut's compressive strength is 250 N/mm² and its Young's modulus $E = 100\ 000$ N/mm², what vertical load P is likely to cause collapse if the pivots are (a) well-fitting, (b) worn and loose?

Force in strut $= P/\sin 55°$ by triangle of forces. First, check for compression failure: $P_y = 250 \times 12 \times 7 = 21\ 000$ N, therefore $P = 17\ 200$ N.

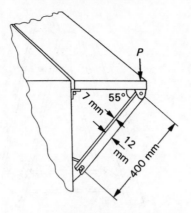

Figure 2.18

(a) Strut can buckle in two directions: in 12 mm direction, hinged conditions, $I = 7 \times 12^3/12$, in 7 mm direction, fixed conditions, $I = 12 \times 7^3/12$

$$P_{ce} = \frac{\pi^2 \times 100\ 000 \times 7 \times 12^3}{12 \times 400^2} \qquad \text{or} \qquad \frac{4\pi^2 \times 100\ 000 \times 12 \times 7^3}{12 \times 400^2}$$

$$= 6218\ N \qquad\qquad \text{or } 8463\ N$$

$P = 6218 \sin 55° = 5093$ N. Short strut treatment, $P = 1/(1/5\ 093 + 1/17\ 200)$ $= 3930$ N.

(b) If pivots are worn and loose, make worst assumption that hinged conditions apply in the 7 mm direction; strength is one-quarter of the fixed–fixed condition; $P_{ce} = 8463/4$ N. $P = 1733$ N (elastic). Short strut treatment, $P = 1574$ N.

Example 2.10

Figure 2.19 shows a frame with legs made from steel tubing of 25 mm outside diameter, 1 mm thickness. Neglecting any stiffness due to the tie-bars at the base, what point load at the top is likely to cause collapse? Let the yield point of the material be 200 N/mm² and its Young's modulus 200 000 N/mm².

The worst case is with load W applied over one pair of legs. If the load is W, force in one tube = $\frac{1}{2}W/\sin 75°$.

$$\text{Failing force in compression} = \frac{200\ \pi(25 - 23^2)}{4}$$

$$= 15\ 080\ N \text{ per tube}$$

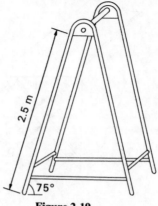

Figure 2.19

Fixed–free buckling case

$$\text{force} = \frac{1}{4}\pi^2 \frac{EI}{L^2} = \frac{\frac{1}{4}\pi^2 \times 200\ 000 \times \pi(25^4 - 23^4)}{64 \times 2500^2}$$

$$= 429\ \text{N} \qquad \text{combined failure } 417\ \text{N}$$

$$W = 2 \times 417 \sin 75° = 806\ \text{N}$$

Problem 2.9

A certain aluminium scaffold pipe is of 50 mm outside diameter, 4 mm thick wall. If its Young's modulus = 72 000 N/mm², find the force at which a 4 m length would buckle elastically under pinned end conditions.

Ans. 6840 N.

Problem 2.10

A square table has four legs, 1.1 m long, of steel tubing 25 mm square externally, 1.2 mm wall thickness. At what load would the table collapse if the load is (a) central, (b) midway between the centre and one leg. Make reasonable assumptions in case b. Take $E = 200\ 000$, $\sigma_y = 250\ \text{N/mm}^2$.
Ans. 15 kN (approx), 7 kN (approx) (leg nearest load gets approximately 50 per cent of force).

2.4 SHEAR

2.4.1 General Remarks

Shearing, that is, cutting with shears, involves a transverse force on the material being cut; in engineering we extend the meaning of shear to mean the stress and strain due to a transverse force without necessarily cutting the material. In figure 2.20a the material between the shearing blades is under shear stress and in a state of shear strain; in addition there is a certain amount of bending. Shear *failure* is by sliding action (figure 2.20b).

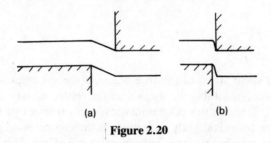

(a)   (b)

Figure 2.20

For shear at its simplest, consider a link containing a vibration-insulating rubber–metal sandwich used in a motor-car gearshift. A better design would use two sandwiches to improve its stability in compression. Here we consider just one (figure 2.21). The rubber block is of length a, width b normal to the paper. The force P causes it to deflect an amount Δ (figure 2.22).

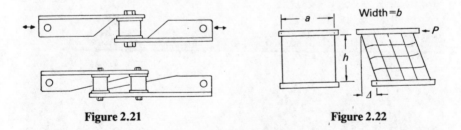

Figure 2.21   **Figure 2.22**

By definition the shear stress = force/area

$$\tau = \frac{P}{ab}$$

By definition the shear strain = deflection/length

$$\phi = \frac{\Delta}{h}$$

By definition stress/strain = modulus of rigidity

$$G = \frac{\tau}{\phi}$$

Hence

$$\tau = G\phi = \frac{P}{ab} = \frac{G\Delta}{h} \tag{2.14a}$$

Hence

$$\Delta = \frac{Ph}{abG} \tag{2.14b}$$

There can be no shear stress at the free edge, so the shear must fade away towards the surface; inevitably the shear near the centre must be greater than average. Figure 2.22 shows this diagrammatically, confirmed by a foam model in figure 2.23. The foam is slightly unrealistic at the compression corners due to buckling of the membranes which exaggerates the effect. The elements in shear are obviously lengthened along one diagonal and shortened along the other, showing that pure shear is accompanied by tension and compression at 45°.

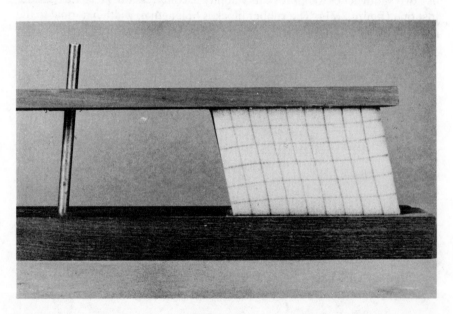

Figure 2.23

2.4.2 Torsion

Torsion of round shafts is the commonest case of shear. Figure 2.24 shows a solid shaft of length L, radius R, with lines and hoops drawn on the surface. When twisted through an angle θ, the hoops remain circular but the lines become helices. The shear strain is, by analogy with figure 2.21, displacement ÷ length, that is, $R\theta/L$.

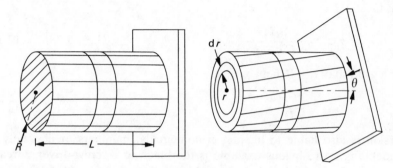

Figure 2.24

To find the torque needed to maintain the elastic twist θ, we consider the shaft as built up from thin tubes of radius r, wall thickness dr. At radius r, the shear displacement is less than at the surface; it is $r\theta$, therefore the shear strain in the elemental tube is $r\theta/L$, hence the stress must be

$$\tau = \frac{Gr\theta}{L} \tag{2.15}$$

The force on the tube = stress × area = $Gr\theta 2\pi r \mathrm{d}r/L$ and the moment of this force about the axis = force × r. Thus the total moment about the axis, that is, the torque = $G\theta/L\int_0^R 2\pi r^3 \,\mathrm{d}r$. The integral is the polar moment of the shaft, generally called J so the torque, T, is given by

$$T = \frac{GJ\theta}{L} \tag{2.16}$$

If we have a hollow shaft, inner radius R_i, all the equations have the same form, the only difference being the limits of integration over which J is determined, so equations 2.15 and 2.16 still apply, using $\int_{R_i}^R 2\pi r^3 \,\mathrm{d}r$.

We thus arrive at a general torsion equation, similar to the general bending equation

$$\frac{T}{J} = \frac{\tau}{r} = \frac{G\theta}{L} \tag{2.17}$$

J is usually expressed in diameter terms since this is the most convenient engineering dimension for shafts, so for a shaft of outer diameter D, inner diameter d

$$J = \frac{\pi(D^4 - d^4)}{32} \tag{2.18}$$

Equation 2.17 gives the stress at any radius of a circular shaft. Mostly we require the highest stress which is found at the outer surface, radius $= D/2$. This is

$$\tau_{max} = \frac{16TD}{\pi(D^4 - d^4)} \tag{2.19}$$

For a solid shaft $d = 0$, hence

$$\tau_{max} = \frac{16T}{\pi D^3} \tag{2.20}$$

Torsion is also liable to happen in non-circular members, deliberately or incidentally. An obvious example is the blade of a screw-driver. Greatly enlarged versions of flat or cross-point screw-drivers are used for drive shafts in rolling mills. The most common shape is the rectangle. The shear strain pattern is shown in figure 2.25. The highest shear stress in a rectangular bar of width b, thickness t under torque T is

$$\tau_{max} = \frac{T(3b + 1.8t)}{b^2t^2} \tag{2.21}$$

The twist θ of a member of length $L \gg b$ is

$$\frac{TL}{G \times \frac{1}{3}bt^3(1 - 0.63\,t/b + 0.053\,t^5/b^5)} \tag{2.22}$$

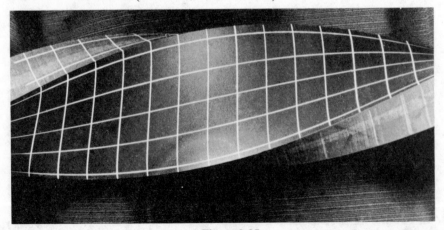

Figure 2.25

while for a very short member the ordinary polar moment is more appropriate provided the ends are unable to develop the warping seen in figure 2.25. This restraint is lost within a length of b to $2b$.[4] For further details reference 5 is suggested.

The stress in longish members given by equation 2.21 is that at the centre of the longer side. Elsewhere on the surface the stress is smaller, approximately inversely as the distance from the centre.

Hollow rectangular sections are widely used in structural steelwork and also in machine design as chassis members, levers, etc. Their properties in torsion are roughly similar to those of a circular pipe of the same mean size and weight. More accurate values are given in appendix A.14 listing standard sections to ISO 657 part 14. For the designer's convenience the following formulae have been derived

$$\text{Twist, } \theta \text{ radians} = \frac{TL}{GK} \quad \text{where } K \approx \frac{2B^2H^2t}{B+H+7t} \tag{2.23}$$

$$\text{Shear stress, } \tau_{max} \approx \frac{T(B+H)}{3.85\,K} \quad \text{within } \pm 4 \text{ per cent} \tag{2.24}$$

or more accurately

$$\tau_{max} \approx \frac{T(B+H)}{3.7\,(H/B)^{1/8}K} \tag{2.25}$$

where H and B are the external dimensions taking $H > B$, t being the wall thickness. These expressions agree well with the listed values which in turn are claimed to represent the standard sections with due allowance for stresses at the corner radii.

The standard expression for rectangular hollow sections does not claim to take corner effects into account. It may be found in reference 5, p. 293 and in various civil engineering manuals.

The most common torsional component other than a simple shaft is the helical spring. To show that this is essentially a coiled-up shaft in torsion, consider the end view, figure 2.26b. At any point on the spring except the piece C, the end force P exerts a torque on the spring material, of magnitude $PD/2$. If the helix is relatively steep a small correction is needed, see appendix A.7.

Tension and compression springs are treated similarly for calculations; tension springs may have a weak point at the end-fastening. Compression springs have an additional safety aspect that provided they are guided, for example, in a hole, fracture can still leave them functioning though with reduced force. Against that, if they are not guided they can buckle like a strut

$$\text{the equivalent } I \text{ value} = d^4h/73ND \tag{2.26}$$

where h is the net height of the spring in the working state (see appendix A.7). If the working range is large, h and P should be checked at several points.

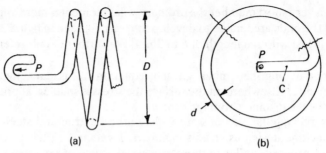

(a)

(b)

Figure 2.26

The deflection of a helical spring is readily derived by energy. Under gradual loading in elastic conditions, the mean force = $P/2$ if the final force is P, thus the work done is $P\Delta/2$. The strain energy in a rod of length L, diameter d under a torque T is $T\theta/2$, by a similar argument. From the torsion equation $T/J = \tau/r = G\theta/L$; $\theta = TL/GJ$, therefore energy = $T^2L/2GJ$.

Equating the work done to the strain energy and noting that $T = PD/2$

$$\frac{P\Delta}{2} = \left(\frac{PD}{2}\right)^2 \frac{L}{2GJ}$$

In a spring of mean coil diameter D, with N turns of wire, $L = \pi ND \sec \alpha$; take $\sec \alpha \approx 1$. For a round wire, $J = \pi d^4/32$. Therefore

$$\Delta = \frac{PD^2}{4} \frac{\pi ND}{G\pi d^4/32} = \frac{8PND^3}{Gd^4} \tag{2.27}$$

The stress is subject to a slight concentration due to the curvature. The basic torsional stress is $Tr/J = (PD/2)\pi d^3/16 = 8PD/\pi d^3$.

The obvious components are due to (a) the inner edge being shorter than L in the ratio $1 - d/D$ so that the twist is steeper at the inside, (b) the direct shear $P/(\pi d^2/4)$ which adds to the torsional shear, increasing it in the ratio $1 + d/2D$. There is also a displacement of the neutral axis, tending to offset this effect.

A common procedure is to use the 'spring constant' $C = D/d$, then

$$\tau = \frac{8PD(C + 0.2)}{(C - 1)\pi d^3} \tag{2.28}$$

It is rare to use C values < 6 or > 40.

Working stresses in springs can be high in many applications, because the load is very predictable provided that the travel is limited by some means. In compression springs the design is normally such that overstress is prevented by letting the coils come together at full load. It is common practice to make the springs originally too long, then press them down to solid in order to yield the material locally. This leaves a favourable residual stress to resist fatigue; it also has the additional advantage that in the course of time this stress creeps out so that the spring tends to lengthen for this reason, which offsets the natural tendency for creep-like shortening under load.

Where there is a shortage of space, rectangular wire can be used for springs. From equations 2.21 and 2.22 it is possible to develop equivalent design formulae, substituting $b^2t^2/(3b + 1.8t)$ for $\pi d^3/16$ and $(1 - 0.63t/b + 0.053t^5/b^5)bt^3/3$ for J, in the stress and deflection expressions respectively. The benefit is often small; it may be better to consider two round wire springs nested concentrically and wound in opposite helices to avoid tangling.

If we wish to take the helix angle into account we must note that as the spring deflects the coil diameter changes, increasing with compression and decreasing with extension. We also observe a tendency to rotate, unwinding with compression and tightening up under extension. In some set-ups this rotation is free to take place, in others it is prevented or hindered to some extent (see appendix A.7).

The typical variables in spring design problems are not neatly prepared in terms ready to pop into the equations. The usual limits are

(a) maximum force and stress;
(b) a given minimum force at partial compression a certain distance from maximum;
(c) outside diameter (including provision for spreading during deflection);
(d) maximum solid length.

From a and b we can derive a stiffness, P/Δ, from d we get a maximum for Nd (including end turns, depending on detail design) and from c we can guess a provisional mean diameter D, then a little trial and error soon produces results. It is useful to note that τ varies as d^{-3} but Δ/P as d^{-4}.

It is difficult to give advice about end finish. If space is very restricted, the end turn is made close-coiled and then most of it is ground away. In this arrangement the tapered run-out contributes to the deflection but there is danger of weakness due to heating during grinding; it is safer to keep the grinding away from the active turns. Sometimes it pays to just cut the ends and make helical supports. In large springs the ends are swaged down before forming the spring.

Another item which needs checking and which is widely ignored in textbooks is the surge frequency. A helical spring in any fixed position has a natural frequency of the free coils surging to and fro. The derivation of this frequency is given briefly in appendix A.7. For steel springs $N = 350\ 000\ d/nD^2$ Hz if all measurements are in millimetres, for inch units this converts to $N = 14\ 000\ d/nD^2$ Hz. The units appear to be inconsistent; this is because the factor includes the shear-wave speed.

It is important to avoid resonance between the working frequency and the surge frequency; the damping in a helical spring is small so that excitation should be avoided also at half, one-third, etc., of the surge frequency if feasible.

Reference 6 is suggested as a guide to materials and design stresses, end forms, etc.

Example 2.11

A helical spring is made of steel, $G = 78\ 000\ \text{N/mm}^2$, $D = 50\ \text{mm}$, $d = 6\ \text{mm}$, $n = 7$ turns. Find the stress and deflection when the load is 80 kg.

$$P = 80 \times 9.81\ \text{N} \qquad \tau = 8 \times 80 \times 9.81 \times 50\ \frac{(50/6 + 0.2)}{(50/6 - 1) \times \pi \times 6^3}$$

$$= 548\ \text{N/mm}^2$$

$$\Delta = \frac{8 \times 80 \times 9.81 \times 50^3 \times 7}{(78000 \times 6^4)} = 54.3\ \text{mm}$$

Example 2.12

A phosphor bronze spring is required for a safety valve, to exert a force of 200 N when in working position with the valve closed. The force must increase about 10 per cent when the valve lifts 2 mm. The stress must not exceed 300 N/mm². Specify a suitable spring. $G = 40\ 000\ \text{N/mm}^2$, $E \approx 2.5\ G$.

(1) Select a provisional D/d ratio. As a starting value, try 6; work on a maximum force basis;

(2) From equation 2.28, $300 = 8 \times 220 \times 6d \times 6.2/(5\pi d^3)$; $d^2 = 8 \times 220 \times 6 \times 6.2/5 \times 300\ \pi = 13.89$. This gives $d = 3.72$ mm. Nearest standard wire gauges are (Imperial) 3.7, 4.1 mm. Either modify D/d or choose a larger value to give lower stress.

(3) To find the deflection, note that force $\propto$ deflection; if the deflection at 200 N force is Δ_1, then at 220 N force and 2 mm extra deflection, $(\Delta_1 + 2)/\Delta_1 = 220/200$, $\Delta_1 = 20$ mm. At maximum force of 220 N, deflection $= 22$ mm.

From equation 2.27

$$22 = \frac{8 \times 220 \times n \times (6d)^3}{40\ 000\ d^4}$$

$$\frac{n}{d} = \frac{22 \times 40\ 000}{8 \times 220 \times 6^3} = 2.315$$

If we choose $d = 4.1$ mm, $n = 9.5$ turns.

(4) Spring design: 8 SWG wire (4.1 mm diameter), mean diameter 24.6 mm, 9½ free turns, assume two end turns. Free length $(9½ + 2) \times 4.1 + 22 = 69.15$ mm, say 70 mm, to be installed in 50 mm space when static. Solid length

= free length − 22 mm. It is understood that springs vary slightly, hence space is usually provided, with adjustment facilities.

(5) Check against buckling using equation 2.26. Fixing conditions not stated, could be pinned both ends

$$\frac{\pi^2 EI}{h^2} = \frac{\pi^2 Ed^4 h}{73Dh^2} = \frac{\pi^2 \times 10^5 \times 4.1^4 \times 50}{(73 \times 24.6 \times 9.5 \times 50^2)} = 327 \text{ N}$$

It is advisable to angle-fix one or both ends but not critical since due to the closeness of coils the spring could not buckle far.

2.4.3 Shearing Stress In Beams (Second year work)

If a beam behaved like a pack of cards each card would be a thin beam taking a small fraction of the load. This is the action of multi-leaf (laminated) springs. In most beams we prefer rigidity. If we had a pure couple, rare in practice, it would be enough to clamp the ends of the pack together. Under a transverse force this would not achieve much, reducing the deflection to ¼ (figure 2.27a, b, c). To get the benefit of a whole beam we would glue the cards together, preventing the sliding by establishing a shearing stress.

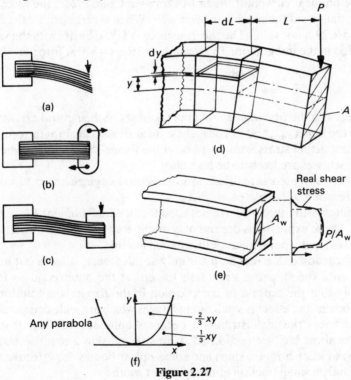

Figure 2.27

To find the shearing stress we take the cantilever, (d). Any layer within the length dL has a bending stress My/I. If the layer is above the neutral axis at a distance y, the tensile stress will be PLy/I at the near face, but $P(L + dL)y/I$ at the far face. Thus the layer experiences a net tensile force of magnitude = stress difference × area; $F = (P\,dLy/I) \times b\,dy$.

The outermost layer is kept in place by the shear stress τ on its lower surface; other layers have shear stresses on both faces so the shearing force on a general layer = area × increment of shear stress, in other words $F = b\,dL\,d\tau$. From these two equations we cancel b and dL, obtaining $-d\tau = Py\,dy/I$

$$\tau = P \int_{y_{max}}^{y} \frac{y\,dy}{I} \tag{2.29}$$

Notice the limits; this is a parabolic equation with maximum of $P\,y^2_{max}/2I$ at $y = 0$. The shear stress is zero at the free surface and builds up to a maximum at the neutral axis.

In a parabolic distribution the maximum is 1.5 times the average. In case the reader has forgotten this piece of geometry, we include a reminder in figure 2.27f. Therefore the peak shearing stress in a rectangular bar = 1.5 P/A. This is a relatively unimportant result since shear stresses are generally low anyway and the high stress in shear comes where the bending stress is low.

The serious shear stresses arise in I-beams, channels, etc. This is obvious from the physical viewpoint. In an I-beam as in figure 2.27e, the force is fed into the flange via the much narrower web. When b varies with y the simple equation 2.29 is not valid. The shearing force = $(P/I)\int by\,dy$, with the varying b included in the integration. The local shear stress = shear force/local b

$$\tau = \frac{P}{I} \int_{y_{max}}^{y} \frac{by\,dy}{b_y} \tag{2.30}$$

The integral is the first moment of area taken about the neutral axis but only beween the limits y_{max} and y. The highest shear stress is again at the centre but the highest actual stresses may well be at the flange–web junction where the bending stresses are likely to be high also.

A widely accepted simplification in structural engineering is to take the shear stress as load/web area, or P/A_w.

It is noteworthy that we have not used the bending moment PL, only the force P. These expressions depend only on the transverse force, whether due to one load, many point loads or distributed loading.

The shear force does not stop abruptly at the corner; it feeds out into the flanges until the shearing stress falls to zero at the outer edge. It follows inevitably that the tension or compression in the flange is not uniform. In narrow beams the effect is small but in flanges which are wide compared with their thickness the maldistribution becomes quite severe. This feature is known as shear lag, see section 8.2.3. In stressed-skin aircraft wings, fuselages, box girder bridges, ships and some railcar bodies the effective flange width is only a small fraction of the apparent width.

Figure 2.28 shows the shear distribution in a rubber beam ruled with a rectangular grid. Although the lines depart considerably from straight, the tension and compression are seen to increase progressively from the neutral axis outwards despite the S-bend introduced by the shear. Note also how the shear penetrates into the support region. It is rare to get a cantilever as per theory; the effective length is appreciably greater than the protruding length.

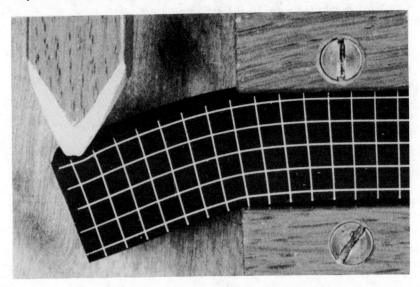

Figure 2.28

In beams the deflection due to shear is rarely important compared with the bending deflections. It is usual to either neglect it or to calculate it on the assumption of uniform shear stress. A truer method is to consider energy. The work done by a load P applied progressively until it reaches its steady deflection Δ is $P\Delta/2$. The energy in unit volume under a shear stress τ applied progressively is final stress $\times$ strain/2 or $\tau\phi/2$ or $\tau^2/2G$. Thus for the total energy in a body we should take not the mean stress but the rms (root mean square) stress which for a parabola is 1.095 times the mean, hence the shear deflection of a rectangular beam is 1.2 times that based on uniform average shear.

Example 2.13

A universal beam 406×178 mm $\times 74$ kg/m is actually 413 mm high, 180 mm wide. The flange thickness = 16 mm, web 9.7 mm. $I = 27\ 330 \times 10^4 \times$ mm^4 (see tables in appendix A.14). What shear force will produce a shear stress of 150 N/mm^2 at the centre? What will the shear stress be at the web–flange junction, ignoring fillet radii? What shear stress would be obtained by the web area method?

$\int by \, dy$ will be obtained by using area and centroid for web and flange separately

½ web

$$\text{height} = \frac{413}{2} - 16 = 190.5 \text{ mm}$$

$$\text{moment} = 190.5 \times 9.7 \times \frac{190.5}{2} = 176\,008 \text{ mm}^3$$

flange

$$\text{centroid height} = 190.5 + 8 \text{ mm}$$

$$\text{moment} = 180 \times 16 \times 198.5 \; = \mathbin{|} \frac{571\,680 \text{ mm}^3}{747\,688 \text{ mm}^3}$$

From equation 2.30

$$150 = \frac{P}{27\,300 \times 10^4} \times \frac{747\,688}{9.7}$$

$$P = 531\,851 \text{ N}$$

τ at 190.5 mm use $\int by \, dy$ for flange alone

$$\tau = \frac{531\,851 \times 571\,860}{27\,330 \times 10^4 \times 9.7} = 114.7 \text{ N/mm}^2$$

τ by web area

$$\tau = \frac{531\,851}{(413 - 32) \times 9.7} = 144 \text{ N/mm}^2$$

Comment: τ by web area is very similar to τ at centre; $144 \approx 150 \text{ N/mm}^2$.

2.5 THE CONNECTION BETWEEN TENSION AND SHEAR

To convince the reader that a member in simple tension is also in shear, figure 2.29 shows a member in tension on which has been drawn a diamond-shaped element. Putting the piece into tension lengthens one diagonal, obviously. The other diagonal is shortened according to Poisson's ratio. What was a square element in the load-free state is now distorted into a sheared shape as in figure 2.29. To quantify this, consider a sloping plane in a tensile member, at an angle θ to the axis. It may help the imagination if we think of a glued joint.

The normal cross-section of the member $= A$, hence the joint area $= A$ cosec θ. Resolving the force P into two parts, a sliding component $P \cos \theta$ and a separating or tensile component $P \sin \theta$, gives the following stresses on the sloping plane

$$\text{Shear stress } \tau_\theta \quad = \frac{P \cos \theta}{A \text{ cosec } \theta}$$

$$\text{Tensile stress } \sigma_\theta = \frac{P \sin \theta}{A \text{ cosec } \theta}$$

Writing the normal tensile stress in the member as σ, then $\sigma = P/A$. The stresses on the sloping plane become

$$\tau_\theta = \sigma \sin \theta \cos \theta = \tfrac{1}{2} \sigma \sin 2\theta$$

$$\sigma_\theta = \sigma \sin^2 \theta$$

The greatest shear stress $\tau_{max} = \sigma/2$ when $\theta = 45°$ (2.31)

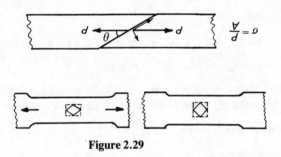

Figure 2.29

2.6 COMBINED TENSILE AND SHEAR LOADS
(Advanced material, may be omitted at first reading)

The most common combinations are two tensions without imposed shear or single tension with shear parallel and normal to the tension. Inclined tension and shear cases are too rare to be considered here and would have to be broken down into orthogonal stresses for ease of analysis. We need go no further than two tensions at right angles and a shear on the same planes. Consider the stresses σ_x and σ_y as shown in figure 2.30, with a shear stress τ. The vertical and horizontal shear stresses co-exist and are equal since shear strain is a matter of the angle of distortion; the material cannot distinguish between a strain as shown at (a) or at (b).

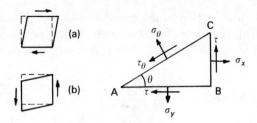

Figure 2.30

Let AC = unit length and let the width normal to the paper be unity (inches or mm as you wish). Then the sloping area = 1, the horizontal area (face AB) = $\cos\theta$, face BC = $\sin\theta$. Resolving normal to AC, the force equation gives

$$\sigma_\theta = BC\,\sigma_x \sin\theta + AB\,\sigma_y \cos\theta - BC\,\tau \cos\theta - AB\tau \sin\theta$$

$$= \sigma_x \sin^2\theta + \sigma_y \cos^2\theta - \tau \sin 2\theta \tag{2.32}$$

Resolving along AC

$$\tau_\theta = BC\,\sigma_x \cos\theta - AB\,\sigma_y \sin\theta + BC\,\tau \sin\theta - AB\tau \cos\theta$$

$$= \tfrac{1}{2}(\sigma_x - \sigma_y)\sin 2\theta - \tau \cos 2\theta \tag{2.33}$$

The highest and lowest tensile stresses are known as the principal stresses. They are free from shear, occur at 90° to each other and are of magnitudes

$$\tfrac{1}{2}(\sigma_x + \sigma_y) \pm [\tfrac{1}{4}(\sigma_x - \sigma_y)^2 + \tau^2]^{1/2} \tag{2.34}$$

The greater stress comes at

$$\theta = \tfrac{1}{2}\arctan\frac{2\tau}{\sigma_y - \sigma_x} \tag{2.35}$$

The greatest $\tau_{max} = [\tfrac{1}{4}(\sigma_x - \sigma_y)^2 + \tau^2]^{1/2}$ \tag{2.36}

The greatest shear stress is not necessarily in the plane which we have been considering but could well be at right angles to the paper. If there is no shear in the third direction, then the greatest shear τ is usually half the greatest tensile stress; we shall call it out-of-plane shear. We are very interested in shear stress because this is what starts the yielding.

To clarify this point, consider the hollow sphere under internal pressure, figure 2.31a. By symmetry, any given imaginary patch extends equally both ways, so the wall is apparently not in shear. The yielding simply takes place out-of-plane as shown in the sectional view. On the other hand when we have tension with compression, for example, as in section 7.10, the shearing tendency is increased as shown in b.

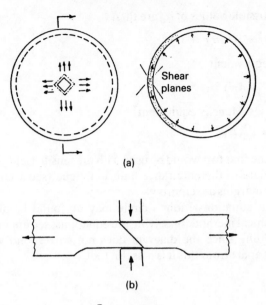

(a)

(b)

Figure 2.31

Many common design cases have shear combined with only one serious tensile stress so that σ_y can be treated as zero. Then the equations reduce to the following

$$\sigma_{max} = \tfrac{1}{2}\sigma + \left(\frac{\sigma^2}{4} + \tau^2\right)^{\tfrac{1}{2}} \tag{2.37}$$

$$\tau_{max} = \left(\frac{\sigma^2}{4} + \tau^2\right)^{\tfrac{1}{2}} \tag{2.38}$$

The most common case is bending combined with torsion. In design work we usually calculate with permissible *tensile* stresses rather than shear stresses when both are present, therefore we convert the stress system into an equivalent pure tensile stress σ_e. There are three variants of σ_e, to match the possible failure modes. If we regard the onset of yield as functional failure, we test for the maximum shear stress. Thus $\sigma_e = 2\tau_{max}$ and must be below yield point, but also σ_{max} must be below yield or we get out-of-plane shear yielding. The tensile stress σ_{max} is also relevant in its own right when using brittle materials that break with negligible plastic flow. Finally when studying long-term fatigue endurance we use a criterion based on strain energy. For the case under discussion here, this boils down to a fairly simple expression (see equation 2.40).

The three possible values of σ_e are then

$$\tfrac{1}{2}\sigma + [(\tfrac{1}{2}\sigma)^2 + \tau^2]^{1/2} \qquad\qquad (2.37\ again)$$

or the Tresca equivalent

$$(\sigma^2 + 4\tau^2)^{1/2} \qquad\qquad (2.39)$$

or the Von Mises–Hencky equivalent

$$(\sigma^2 + 3\tau^2)^{1/2} \qquad\qquad (2.40)$$

the greater of the first two would be used with the tensile yield point, while the third would relate to the endurance limit in fatigue (see section 3.4.1) with suitable safety margins as required.

For the long equations Mohr's circle may be found helpful. This is an ingenious graphical method for solving the equations, see appendix A.11. It is slightly misleading since the diagram does not correspond with the stress directions, but apart from this it is a useful tool.

2.7 MULTIPLE SHEAR

Many students find difficulty in visualising the flow of force in an assembly of several parts. We often use a clevis or blade-and-fork arrangement in which the pins, bolts or rivets resist the load on two planes simultaneously. In some aircraft joints this is taken even further, with three or four shear planes. In some cases the shear is so well subdivided that the limiting feature arises at the contact faces, the local compressive (bearing) stress becoming excessive (see section 8.7.2). In a clevis arrangement the symmetry of the joint allows us to assume equal load sharing in any one pin; when two or more pins are used, the arguments regarding load sharing are too long to be debated here. In designing for steady loads with ductile materials, and using compact grouping, we rely on equalisation by local yielding long before the onset of failure and assume equal load sharing. We also generally assume even distribution across the plates of the joint. Under fatigue loading these assumptions are not valid; we have to consider shear distribution as discussed in section 2.4.3. Even more, we must consider stress concentration around the hole (see chapter 4) and fretting.

For clarification, see example 2.15 (below) and problems 2.16, 2.17.

Problem 2.11

A rubber block similar to figure 2.21 is required to have a stiffness in shear of 40 N/mm. The grade of rubber to be used has a modulus of rigidity $G = 0.6$

N/mm^2. (You may find it convenient to think of the duty as a force of 40 N when the lateral deflection is 1 mm.) Find the thickness required if the shear area measures (a) 30 mm × 30 mm, (b) 40 mm × 40 mm, (c) 60 mm × 40 mm.

Ans. 13.5, 24, 36 mm.

Problem 2.12

A rubber mounting is shown in figure 2.32. The central member deflects 6 mm when carrying a load of 50 kg. What is the modulus of rigidity of the rubber? (Taken from an actual project.)

Ans. 0.54 N/mm^2.

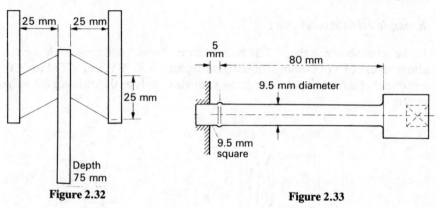

Figure 2.32   **Figure 2.33**

Problem 2.13

Figure 2.33 shows an extension rod for a torque spanner. Find the shear stresses in the square and round portions when a torque of 50 N m is applied to it and *estimate* the twist. Take $G = 80\ 000\ N/mm^2$

Ans. 280, 297 N/mm^2, 3.6° (¹⁄₁₆ radian).

Problem 2.14

The steel torsion spring used in a 400 day clock is 1.2 mm wide, 0.15 mm thick, 90 mm long. What shear stress is produced in this spring if it twists 1.5 radians from the straight state? Take G as 80 000 N/m^2.

Ans. ≈ 200 N/mm^2.

Problem 2.15

Design a helical compression spring for the following duties, in steel, $G = 80\ 000\ N/mm^2$. Solid length ≤ 20 mm, mean diameter $D = 25$ mm, helix angle may be ignored. Working force = 360 N ± 3 per cent when length = 25 mm (valve spring, valve fully open)

Stress at working force ≤ 700 N/mm²
Stress when compressed solid ≤ 880 N/mm²
Force when length is 32 mm ≥ 200 N   (valve closed)

Since D/d is not known, make first attempt ignoring correction, or guess 6 to 8.

Ans. d = 3.4 mm, 4 free turns + 2 end turns, free length = 42 mm.

This question uses values from an inexpensive but highly reliable automotive valve spring believed to be made of chrome–vanadium steel.

Example 2.14 (Second year)

In the lever shown in figure 2.34 find the force P which will cause yielding at a shear stress of 115 N/mm². Investigate regions XX, YY and ZZ. For YY, assume that the torsional shear stress varies inversely as the distance from the centre.

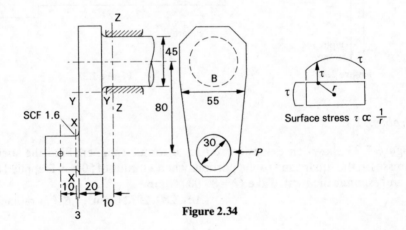

Figure 2.34

XX:

$$\text{area} = \pi \times \frac{30^2}{4} = 707 \text{ mm}^2$$

$$\text{shear stress} \approx 1.5\, \frac{P}{A} \ (\text{section } 2.4.3) = 2.1\,P \times 10^{-3} \ {}^*$$

$$\text{Bending stress } \frac{My}{I} = 10P \times \frac{32}{\pi d^3} = 3.77P \times 10^{-3}$$

* Actually for round bars peak shear stress is 4/3 times average transverse shear stress.

Combined effect not required since shear high where bending low

$$\text{Max. } \sigma = \text{SCF} \times 3.77 \times 10^{-3}P = 0.006\,P$$

(SCF means stress concentration factor)

YY:

$$\text{area} = 20 \times 55 = 1100 \text{ mm}^2$$

$$\text{max. direct shear} = 1.5\frac{P}{A} = 1.36P \times 10^{-3}$$

$$\text{Torque} = 23P \text{ at centre} = 23P\frac{(3 \times 55 + 1.8 \times 20)}{20 \times 55^2} = 3.82P \times 10^{-3}$$

$$\text{Bending moment} = 56P \qquad \sigma = \frac{6M}{bh^2} = \frac{6 \times 56P}{20 \times 55^2} = 5.55P \times 10^{-3}$$

Combined stresses: at centre of long side direct shear and torsional shear are high, bending nil. Equivalent tensile stress for yielding $\sigma_e = 2\tau$

$$\sigma_e = 2\left(\frac{0.003\,82P + P}{20 \times 55}\right) = 2 \times 0.005\,2P = 0.01P$$

$$\text{or } \tau = 0.005\,2P$$

at centre of short side, bending stress $= 0.0055P$ (see above)

$$\text{torsional shear} = 0.003\,82 \times \frac{20}{55} = 0.001\,39P$$

$$\text{direct shear} = \text{ nil at surface}$$

$$\sigma_e = (0.0055^2 + 4 \times 0.001\,39^2)^{1/2}P = 0.006\,2P$$

Intermediate positions should be investigated too, but are omitted here.

ZZ:

$$\text{area} = \pi \times \frac{45^2}{4} = 1590 \text{ mm}^2$$

$$\text{max. direct shear } 1.5\frac{P}{A} = 0.94P \times 10^{-3}$$

$$\text{Torque} = 80P \quad \text{torsional shear} = \frac{16 \times 80P}{\pi \times 45^3} = 4.47P \times 10^{-3}$$

$$\text{Bending moment} = 43P \qquad \sigma = \frac{32 \times 43P}{\pi \times 45^3} = 4.8P \times 10^{-3}$$

Combined stresses:

at B, add direct and torsional shear, no bending. $\sigma_e = 0.011P$

at C, direct shear $= 0$, $\sigma_e = (4.8^2 + 4 \times 4.47^2)^{\frac{1}{2}} \times 10^{-3}P = 0.011P$.

Answer: highest $\sigma_e = 0.011P$, hence highest $\tau = 0.005\,55P$. For 115 N/mm²,
$P = 20$ kN.

Comment: sections below YY where arm is narrower may need investigating.

Example 2.15

The joint in figure 2.35a uses a pin of 10 mm shank diameter, that in figure
2.35b has two pins each of 8 mm shank diameter. Find the nominal stresses in
the pins and in the plates when a load P of 8 kN is applied.

(a) Pin:

$$\text{shear force} = 4 \text{ kN per shear face (ignoring friction)}$$

$$\text{area} = \frac{100\pi}{4} \text{ mm}^2$$

$$\tau = \frac{4000}{\text{area}} = 51 \text{ N/mm}^2$$

Central blade:

$$\text{net width} = 22 - 10 \text{ mm} \qquad \text{area} = 11\,(22 - 10) = 132 \text{ mm}^2$$

$$\sigma = \frac{8000}{132} = 60.6 \text{ N/mm}^2$$

Side cheeks:

$$\text{area of both together} = 12\,(22 - 10) = 144 \text{ mm}^2$$

$$\sigma = 55.6 \text{ N/mm}^2$$

(b) Pins: four faces sharing

$$\text{area of one face} = \frac{\pi \times 8^2}{4} \text{ mm}^2 \qquad \tau = \frac{2000}{\text{one area}} = 39.8 \text{ N/mm}^2$$

Central blade:

$$\text{net area} = 11(22 - 8) = 154 \text{ mm}^2 \quad \sigma = \frac{8000}{154} = 52 \text{ N/mm}^2, \text{ at X–X}$$

Side cheeks:

$$\text{net area} = 12(22 - 8) = 168 \text{ mm}^2 \quad \sigma = \frac{8000}{168} = 47.6 \text{ N/mm}^2,$$
$$\text{at Y–Y}$$

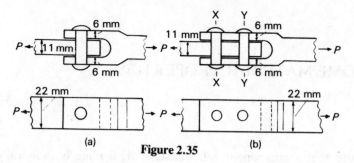

Figure 2.35

Problem 2.16

Find the nominal shear stresses in the pins of the four joints shown in figure 2.36. All pins are of 6 mm shank diameter.

Ans. 17.7, 17.7, 17.7, (a,b,c are equivalent), 177, 133 N/mm²
(d, by taking moments).

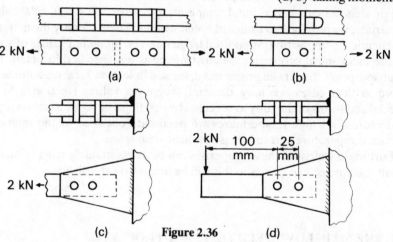

Figure 2.36

Problem 2.17

A boat propeller shaft is fitted with a hard brass shear pin, of failing stress 170 N/mm². If the pin is of 3.5 mm diameter and the shaft 20 mm diameter (figure 2.37), what shaft torque is needed to shear the pin?

Ans. 32 N m.

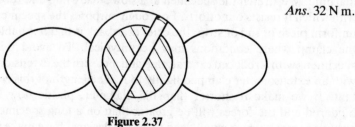

Figure 2.37

3 SOME MATERIAL PROPERTIES

To ensure that a component will withstand its loading economically, we ideally should make several prototypes and test them to failure, thus finding the lightest or cheapest version. This is very slow and extravagant; we rely on calculations for stress and on material data obtained from standard specimens. How do these relate to real needs in terms of size, shape, time factor, temperature, environment and (particularly) load fluctuation? What adjustments are needed to give valid predictions of performance?

For steady loads and normal temperatures, when using well-established materials, we purchase to standard specifications. These identify the material by chemical composition, supported by guaranteed minimum values of some of the mechanical properties. Microstructure is not specified in detail. For many purposes the tensile properties discussed in section 3.2 are adequate; in more critical cases we may demand toughness values (section 3.5) or specialised forms of ductility as revealed by bending a bar over a certain angle and radius. At high temperatures we need data obtained at the intended service temperature and covering a realistic time-span.

Fortunately most of these properties can be assessed safely from relatively small specimens; exceptions to this will be discussed extensively.

3.1 THE MAIN LOW SPEED TESTS AND PROPERTIES

The most basic test is tensile, carried out on smooth specimens of around 10 to 20 mm diameter, in air, at room temperature and lasting several minutes. The machine used is basically a screw or hydraulic jack fixed to a weighing machine. We stretch the specimen at a slow steady rate and record the force with which it resists our efforts. For rough purposes the specimen can be any uniform piece of rod or strip; the chief drawback is that it is liable to break at the clamp where conditions may be untypical. To avoid this, we make specimens with a reduced central zone and measure the extension of this zone with an extensometer clamped lightly to it. The length of this zone is important. If we make it shorter than the diameter, plastic flow is somewhat hindered and the forces will be greater than on a long specimen, giving an optimistic result which would lead to unsafe designs. If we make it very short, like a groove, there will be concentration of stress; local plastic flow will start too early but after that the strength will appear high. A usual length is $4\sqrt{A}$ where A is the cross-sectional area; this is about the shortest length for uninhibited plastic flow. Figure 3.1 shows schematically how a tensile test is laid out, together with some typical results.

The upper curve is typical of steels; the uniformly sloping part is the elastic extension. At P_1 something happens in the material: some of the internal forces are overcome and internal movement begins. Sometimes the movement is visible on the surface, producing Lüder's lines at 45° to the axis. Further extension starts off additional lines. At some stage, the extension localises itself and a thin neck forms in the specimen. This stage comes approximately at the point where the curve turns downwards. There is a balance between increasing true strength per unit neck area and decrease of this minimum area. This action is of interest in forming processes such as tube-making or wire-drawing; in design problems we would normally avoid getting beyond the yield point P_1. We do however take some notice of the highest force stood by the specimen and the amount of stretch at that point; this shows the further safety margin between first onset of flow and actual fracture.

The reduction of area and the true stress at fracture, though mainly of interest in metal-working, give some idea of the reserve of strength at grooves and notches of various kinds. The ability to even out the stresses by local yielding is very important.

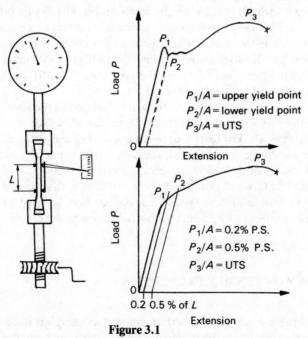

Figure 3.1

The slope of the steep curve could have been predicted from equation 2.2, or alternatively the observed slope can be used to calculate E. The student is warned that in some routine tests the extensometer is omitted, the extension recorded is that between the machine cross-heads and therefore includes the whole specimen, the grips, settlement at the wedges, etc.

While we are on the steep curve we can interrupt the process; if we wind the machine back slowly, the force goes back almost exactly along the same line. Students should be encouraged to do this when testing, as a check on friction, slippage at the extensometer, etc. If we perform this off-loading after yield has started we get a similar elastic recovery but with a slight loop-formation as shown in the upper diagram. Whenever the part is under elastic extension the diameter decreases slightly; likewise with rectangular specimens. For a width B the contraction δB is given by

$$\frac{\delta B}{B} = \frac{\nu \Delta}{L} \tag{3.1}$$

This equation is the definition of ν, Poisson's ratio. For the majority of metals it lies between 0.25 and 0.35. The value of Young's modulus includes this effect; in some cases which we shall meet later this contraction is inhibited and we must use a modified E value. In the constant-stress plastic range, ν becomes 0.5 which implies constant density. In a unit cube, the two lateral contractions would then fully compensate for the extension.

Many materials including some strong steels do not show a distinct yield point, giving a smooth curve as in the lower graph. For design purposes it is difficult to decide where permanent deformation starts since results would vary with the sensitivity of the machine, so we create a definite point by drawing a line parallel to the original elastic line offset by a distance representing 0.1, 0.2 or 0.5 per cent of L (L is called the gauge length). Such lines meet the curve at the 0.1 per cent proof stress, etc.

The important points in tensile testing are direction of the specimen axis in relation to the product and a sufficiently large size to include a large number of individual crystals. For instance, rolled plate can have different properties lengthwise, widthwise and in the thickness direction, especially with regard to slag inclusions. Castings will vary according to the direction of solidification. The grain size aspect will not usually be a problem with standard specimens but can be so with miniature specimens, such as may be used when investigating the properties of fragments from a failed component.

3.2 THEORY OF TENSILE PROPERTIES

When breaking a component a certain amount of energy is used to produce fresh surface, overcoming the molecular forces. If we calculate the work from surface energy data or from the latent heat of evaporation, it works out to be much higher than the observed energy. The stresses to account for the energy would be of the order of $E/8$ or $E/10$, that is, enough to produce elastic extensions of over 10 per cent. Even in linearly grown whisker crystals, carbon fibres, etc., we cannot reach strengths much over $E/60$.

Griffith [7] was led to assume a plentiful supply of defects or misfits in the material, ready to move along or join up. Thanks to electron micrography such defects (dislocations) have been observed. They take various forms, notably in twist which is difficult to describe. For a simplified picture we can take figure 3.2. The dislocation A is able to move along under a stress which only has to affect a few links at a time since some are already stretched by the 10 per cent required in theory; it would then move indefinitely until stopped by another distortion, for example, an alloying element of different atom spacing (B) or a grain boundary.

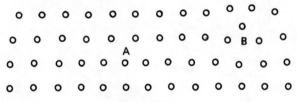

Figure 3.2

This motion is essentially a shear action and could well be independent of any tension either along the plane or normal to it, which ties up well with the observed 45° yield angle (equation 2.31). If it is true that tension makes no difference as such, then we would expect the same yield point in compression as in tension since equation 2.28 is the same regardless of sign. This too is found to be the case, with ductile metals. In relatively brittle materials the yield is not clear; fracture is at 60 to 90° to the load line.

The torsional yield point is harder to observe than the tensile one since the stress is not uniform; however it is found to be about half the tensile yield point; this also agrees with our picture.

We may expect crystal size to affect yield point but not the eventual break when the dislocations have moved as far as they can. The first point is observed to be valid. Large single crystals of pure metals have remarkably low yield points; high-yielding materials have fine grain size, obtained by chilling when casting, by cold-working or by heat treatment involving structural changes. The second point is also largely confirmed if we look at the true stress at fracture, taking account of the reduction of area.

Finally if we had completely symmetrical triaxial tension, the dislocations would not know which way to go. As Griffith predicted, the presence of triaxial tension encourages brittle fracture at the expense of yielding; note that fully symmetrical triaxial tension is rare but partial triaxiality is quite common at the base of grooves.

3.3 LOCAL YIELDING AND RESIDUAL STRESSES

This subject has an important bearing on fatigue behaviour, as we shall see shortly. In any non-uniform stress field in a ductile material it is possible to cause local yielding in tension, compression or shear situations, though presumably the micromechanism is always shear. When the load causing the stress is removed, the structure tries to recover elastically but the locally altered regions no longer fit into their previous places.

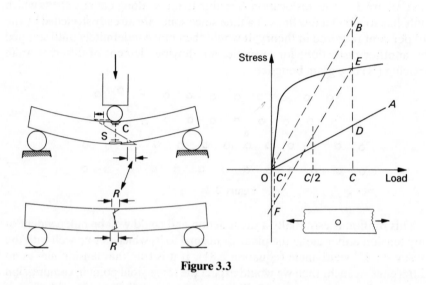

Figure 3.3

The beam in figure 3.3, originally straight, is loaded so much that a small region C yields by shortening and a small region S is stretched. A graph of the stresses is shown superimposed on the beam. When the load is removed the beam does not straighten out completely because the distorted regions go into reversed stresses R thus opposing the full recovery. This reversed stress R is one form of residual stress. It is roughly of the same magnitude as the missing portion in the upper picture.

If the yielding is very localised as for example at the sides of a small hole in a tension member, we can attempt to quantify the residual stress. The general stress follows line OA as we increase the load. The local stress starts off along OB but when it meets the yield curve for the material it is obliged to follow this. At a load C, the general stress is D, the local stress E. If we now remove the load, the general stress returns towards O; the stress in the region of stress concentration immediately gets below yield and therefore cannot follow the plastic flow curve but behaves elastically. The stress concentration factor will be altered by the shape change but the general effect will be something like line EF more or less parallel to BO. When the external load is zero there will be a small internal load C'.

The important aspect is the behaviour at future loadings. Supposing we apply a load $C/2$. The general stress will be about $D/2$ but the local stress will be much smaller than either $B/2$ or $E/2$ as may be seen from the graph. The residual stress is beneficial for all future loads less than C but in the same direction. On the other hand a reverse load is aggravated by the residual stress, in many cases giving reverse yielding. Repeated reverse yielding work-hardens the material; if the loading is not too severe then yielding will stop because the new yield point is higher than the stresses; if the loading is too severe for this, cracking will start and may eventually lead to fracture. Cracking does not always produce fracture; cracks may relieve the stresses by redirecting the lines of force, or may simply stop when they penetrate into a low-stress region.

These opposing trends are frequently misunderstood which is unfortunate because they have an important bearing on the validity of fatigue test procedures if we accept that stress concentrations within the material behave similarly to those resulting from external shape.

Practical evidence for the powerful effects of residual stress due to prior loading is the long-established use of autofrettage in gun-making. This involves giving the gun a hydraulic loading in excess of the working pressures and is found to prolong the life under repeated firing. (See section 10.1 and appendix A.3.)

3.4 FATIGUE LOADING AND FATIGUE TESTS

Sometimes the impression is given that fatigue is a kind of disease that strikes without visible reason. This is nonsense; the designer knows that high stresses repeated many times can cause failure. Designing against fatigue is fundamentally the same as designing against immediate failure; it is a case of making things strong enough. It is however much more difficult to design economically against fatigue than against static failure because

(1) the local stresses are much more important than in static work;
(2) the material properties are less easily established;
(3) success or failure are not immediately apparent.

To describe the types of duty let us consider a circular saw bench for a busy joinery works. Figure 3.4 shows a frame F, saw spindle S, belt drive B and motor M. For simplicity we shall only look at points P on the two shafts. The belt tension causes bending moments. As the shafts rotate, parts of the surface come alternately under tensile and compressive stress, proportional

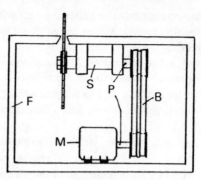

Figure 3.4

to the belt tension. This tension is by no means constant. Its lowest working value is when the saw is idling round between jobs; during cutting it is at something like the full power value but when we first switch on, the motor has to accelerate the big blade and may exert a starting torque up to three times normal, so the peak belt tension will be increased considerably above normal. The shafts are also in torsional shear stress. This is steady in the sense that it does not reverse during every rotation; yet it will have a slight ripple due to individual tooth loads. It will also vary from zero to a peak value every time we switch on and during normal working it will cycle between idling and cutting values.

In designing we have to consider a life of say 10 000 hours. This will include many millions of normal stress cycles, perhaps 10 000 peak torques and 50 000 peak bending cycles (at each switch-on there may be several revolutions at high belt tension as the saw speeds up). Presumably we can forget about corrosion or abnormal temperatures. The biggest problem is testing; if we build a prototype we should like to test it fully in 100 hours rather than 10 000.

The problem needs breaking down into its parts: the fully reversed bending stress, the addition of a steady stress, the spectrum of high and low loads, the sequence in which they occur and any corrections for the effect of speeded-up tests.

3.4.1 Fully Reversed Stresses, Long and Short-life Design

Most of the duties in the example given above fall into the so-called high-cycle region of well over 10^4 cycles of required survival; this duty is far commoner than the low-cycle region examples of which are aircraft landing gear, pressure vessels, etc. Accordingly the high cycle region is much more fully documented; the low-cycle region where we can use higher stresses and hence smaller, lighter components is still being explored. The most popular high-cycle test-pieces are shown in figure 3.5. The basic procedure consists of making a number of identical specimens and running each one under a different bending moment until it breaks. The number of reversals at break are plotted logarithmically against the stress, giving a σ–N curve.

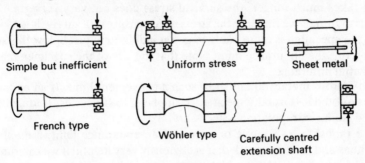

Simple but inefficient   Uniform stress   Sheet metal

French type   Wöhler type   Carefully centred extension shaft

Figure 3.5

The first kind of specimen is relatively inefficient since only a very short length of the material is under the high stress on which the results are based. The greater part of early fatigue data are based on this specimen and show considerable scatter. In the other forms the stress is relatively uniform over a longer length, hence a greater proportion of the specimen is being truly tested. Results are accordingly more consistent.

The kind of σ–N curve generally found is shown in figure 3.6, including test values between yield point and UTS. How do we test above yield point? In

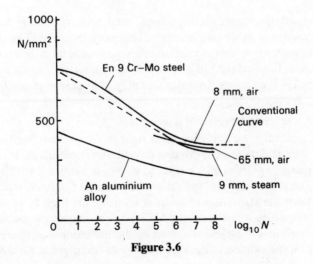

Figure 3.6

fatigue testing it is possible to start rotation first, then put on the load gradually so that the material work-hardens progressively over many reversals. In practice the full load is often there before rotation starts; hence at high stresses many fatigue tests are unrealistic.

We see that between 10^6 cycles which is easy to achieve and 10^8 cycles, which takes much longer, the survival stress does not vary very greatly. In ferrous materials we find that the stress which ensures 10^7 survivals is a kind of limit; the specimen is quite likely to go on indefinitely just below this stress. We call this the endurance limit. Note that the size of the specimen makes a substantial difference.

Non-ferrous metals do not show such a convenient limit. If an endurance limit is quoted it is usually specified as applying to 10^7 cycles. If longer life is needed, some extrapolation is necessary.

To establish a $\sigma{-}N$ curve or even just an endurance limit, a number of specimens and tests are needed; it is obviously very helpful if we can run such tests at high speeds. In the high cycle region about 50 Hz is fairly usual; working at lower or higher speeds seems to give about the same life when measured in cycles. In medium to low-cycle work, speed seems to be rather important. Wöhler [8] found that life at a certain stress level became considerably shorter when testing at 0.3 Hz rather than 1.2 Hz. 108 years later, Musuva and Radon [9] have rediscovered this effect: at 0.25 Hz cracks can grow about three times faster per cycle than at 30 Hz. This time-scale discrepancy is one of the designer's many worries in applying low-cycle fatigue data.

The endurance limit is affected by the environment; as an example the steel referred to in figure 3.6, which is commonly used in a steam environment, shows an endurance limit about 8 per cent lower than in air. [10] This is partly due to the temperature, partly to the corrosive effect of steam in the presence of stress.

An important discrepancy is due to size; small specimens show high endurance limits except in push–pull testing of uniform specimens. The clue to this is the stress gradient; as Griffith said in 1921, we should be interested not only in the stress at the surface but also inside the material, one or more grain diameters below the surface. Initial cracks may well cease to propagate if they

(1) meet a fresh crystal and
(2) have reached a region of lower stress. [11]

This would generally not apply to cracks right across the specimen; only short local ones. The effect of size is discussed further in appendix A.2.

It is not usual to publish $\sigma{-}N$ curves as part of normal material properties, only the static properties and the endurance limit in fully reversed stress. A fair estimate of the stress–life relation at least for the high-cycle end can be obtained by drawing a line from the UTS (that is, one single loading) to the endurance limit at 10^6 cycles, as shown by the dashed line in figure 3.6. This tends to err on the cautious side, thus helping to compensate for size effect.

3.4.2 Partly Reversed and Fluctuating Stresses

In our design example the first complication is the presence of a torque in addition to bending. Though this torque is steady while the bending stress alternates, it raises the peak stress and therefore has some influence on endurance. The fact that it is a shear stress can be overcome by using an equivalent tensile stress, $\tau \times \sqrt{3}$.

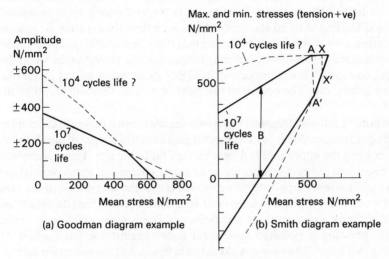

These diagrams refer to material of UTS 800 N/mm², yield 630 N/mm², endurance limit ± 370 N/mm²

Figure 3.7

Mixtures of steady and alternating stress have been researched heavily. The conclusions can be plotted on a Goodman diagram, figure 3.7a. In theory the line could run straight from the endurance limit to the UTS; however we prefer to design within the yield point so we put a kink into the line. To the right of the kink the line follows a law $x + y =$ constant where the constant is given the value of the yield point. Please note that in some older books the alternating stress is plotted as peak-to-peak amplitude.

The same diagram can be used for higher stresses and hence shorter lives, by starting with the alternating stress corresponding to a given expected life from the σ–N curve and drawing a line towards the UTS until it meets the sloping line $x + y =$ yield stress. This approach is suggested in some of the best books on fatigue design but is subject to the uncertainties of low-cycle endurance predictions.

A simple numerical way of expressing the loading is the stress ratio R. It is the ratio of lowest to highest stress. $R = -1$ for full reversal, 0 for repeated loading, etc. A value $R = 0.5$ is a loading varying between half and full value, in the same direction.

We could therefore plot the endurance stress or the stress for a limited life against stress ratio; this is becoming more common. The form which gives the most physical grasp is the Smith diagram, figure 3.7b. This shows the same information as the Goodman diagram but the vertical scale is the total stress. The upper line = mean stress + amplitude, the lower one = mean stress − amplitude, hence the acceptable range for any given mean stress level can be seen directly. One point of importance is the endurance limit in one-way repeated loading with no reversals, this is seen directly as arrow B. Another interesting area is AA′. This indicates that if the material is taken up to yield point it is still capable of withstanding fatigue loading provided the stress does not get too *low*. This is important in highly prestressed bolted assemblies, welded joints, etc. The region to the right of A is mathematical rather than real.†

We note that these diagrams show only tensile mean stresses. Extrapolating to the compressive side is justified although due to the danger of misalignment and buckling the experiments do not go very far to the left. There is considerable direct and indirect evidence that a compressive mean stress permits us to use a higher stress amplitude; the problem is that the magnitude of the mean stress is often not known. A long-established method of strengthening crankshafts, railway axles, etc., at the points of stress concentration is surface rolling; pressing a polished hard steel roller against the shaft while it is rotating in a lathe. The action is indicated in figure 3.8; the indenting action of the roller work-hardens and lengthens the material transversely by compressing it plastically inwards. On release it is in compression. The process must not be overdone otherwise excessive tension below the surface may be produced, or excessive shear stress between layers. When done suitably it can double the effective ± endurance limit. [12]

Other methods of producing a compressive mean stress are shot-peening which is similar to rolling but more adaptable to various shapes, leaving a hammered finish suitable for springs, etc., but not for bearing surfaces, also case-hardening of steels which intrudes extra atoms into the crystal lattice; although care must be taken that the heat treatment does not leave tensile stresses. A variation of this is plasma or ion treatment, a relatively recent invention.

A very important source of residual stress is local yielding as described in section 3.3. Although its main effect is on components with stress raisers, it is also significant in smooth specimens if we consider microscopic inhomo-

---

† This is a consequence of using mean stress as basis. Once yielding has started, the mean stress is no longer an independent variable. The true independent functions are the highest and lowest *strain*; as the upper stress automatically levels off at yield point, the mean stress fixes itself according to the lower strain. If the material can stand the range AA′ it is obvious that it can also stand a smaller range XX′.

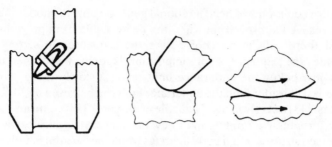

Figure 3.8

geneities. It helps to explain the results of differing loading sequences (appendix A.3). Such residual stresses are unfavourable if tensile; if compressive they may be either good or bad – a compressive residual stress plus a compressive additional load could cause reverse yielding and then leave a tensile residual stress.

Tensile residual stresses are clearly unfavourable to fatigue-loaded components, imposing a mean tensile stress often of unknown magnitude. Welded joints are under considerable tensile stress due to cooling of the final welds. In highly stressed structures these residual stresses may have to be relieved by heat treatment. Grinding of highly stressed shafts is risky since the heat of grinding may leave a thin skin of material in tension as it cools.

3.4.3 Load Spectrum, Design and Data Problems

Returning to our design example, once we have selected the power and speed, we need to decide on permissible stress levels. From the motor and saw blade data we would make a preliminary dynamic calculation to find the peak torque and the number of revolutions over which we expect it to persist. Let us assume that this is 50 000 in a reasonable lifetime. Supposing we use the same material as in figure 3.7, we could allow a stress of 500 N/mm² or thereabouts, depending on the proportions of mean and reversing stresses. However, as this stress is a limited life stress presumably some damage is gradually developing in the material. Is it possible that this will shorten the life at the normal working condition (which could well be required to survive 10^9 cycles)?

In this instance we should probably design to well below the endurance limit even under the peak stress condition; then the only likely dangers would be the works using a weaker material or the customer fitting a stronger motor, or overtightening the belts. In aircraft work where weight is critical and varied loadings are common, we need a rational guide on how to choose the highest safe stresses. A load spectrum will often be available from previous experience and there will be no choice over the sequence. In setting up test programmes the sequence is very important. Running a specimen for many cycles at just below the endurance limit or on a slowly rising programme strengthens it considerably; the endurance limit can be raised by 30 per cent in some cases, [13] notably in highly ductile materials.† This confirms the virtues of gentle, progressive running-in of new machines. It applies particularly to fully reversed stresses. High early loadings would be harmful in fully reversed stress; on the other hand in one-way stressing a high early load can have good or bad effects depending on its direction. Figure 3.9 from Heywood [14] shows the general experience in various practical cases containing stress raisers. In smooth specimens the effect is less pronounced but still present.

Subject to the foregoing, that is, in the absence of either early damage or exceptionally favourable early strengthening, it seems that damage accumulates linearly. If at some stress range 1 the life would be N_1 then every cycle uses up $1/N_1$ of the whole life; similarly for other stresses. So for a variety of stress ranges 1, 2, 3, etc., the damage builds up so that after n_1 cycles at stress 1, plus n_2 cycles at stress 2, etc., the total of damage fractions $= n_1/N_1 + n_2/N_2 + n_3/N_3 \ldots$, written as $\Sigma n/N$, and when this sum reaches 1, that is, 100 per cent, failure is due. This is wrongly known as Miner's law, [15] being first noted by Palmgren, [16] for ball bearings. Miner never claimed it to be a complete law; in fact he designed many of the tests to clarify the effect of loading sequence (see appendix A.3). The damage sum (survival sum would be a better phrase) is often very close to 1 but can range considerably either way under certain circumstances.

The importance to the designer is that he cannot take all test data at face value, some test sequences being too pessimistic while others could lead into danger. When tests on specimens or working parts are carried out without specific trends towards a particular order, it seems safe to assume that $\Sigma n/N > 0.5$. Any rest periods at zero load tend to increase the sum, so that in practice life may well amount to many more cycles than deduced from tests where long rests are not used. Premature failure is more likely if the service includes long dwell periods at full load but even this is not always so. The degree of caution required of the designer in fatigue work must vary from case to case. The wise designer will also employ a fail-safe attitude – in our example he would arrange for the weaker point to come where the pulley is attached to the saw blade shaft, not at the blade end. If anything is wrong with the material, the calculation or the operation, then the first thing to fall off will be the pulley rather than the blade.

† This effect has been investigated substantially, from Gough in 1924 to Miller and Zacharia in 1977 (see appendix A.3). In engineering, similar researches tend to reappear cyclically every 20 to 30 years, mainly because new work does not get into the textbooks and engineers do not read old original papers – there are too many of them.

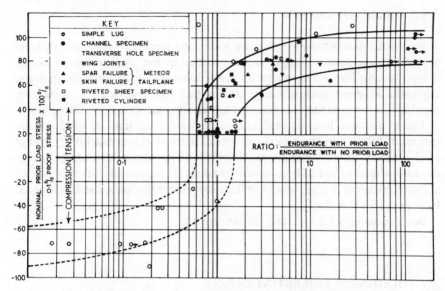

Figure 3.9 Reproduced, by permission of Chapman & Hall Ltd,
from R. B. Heywood, *Designing against Fatigue* (1962)

3.4.4 Torsional Fatigue

Because of their great importance in engines, compressors and many other
products, fatigue properties in torsion alone and combined with bending have
been extensively investigated. In fully alternating torsion the endurance limit
in shear is between 55 and 60 per cent of the tensile endurance limit in
alternating bending. This agrees with the Von Mises–Hencky criterion based
on shear strain energy; the main point is that it is well established experimentally, both in pure torsion and in torsion combined with bending when
both fluctuate together. When there is a steady mean stress there is one series
of results, quoted in reference [13] which finds that the permissible range is
not affected by mean stress on smooth specimens. In most cases there are
stress raisers present which means that the state of stress is complicated by
diagonal tensions, etc., and it is safer to assume that a Goodman or Smith
diagram should be used. It is very difficult to draw a Goodman or Smith
diagram in torsion realistically because the final shear strength is rarely given.
It cannot be deduced from the UTS since the final break area is quite different
from that in a tensile test. Hence for safety the author prefers to pretend that
the torque fluctuates in phase with the bending moment even if it is actually
steady (see section 10.2).

3.4.5 Fatigue and Corrosion

It is obvious that corrosion must affect any long-term test results if only from loss of cross-section. In fatigue there is mutual action: cyclic strain breaks up the protective film which normally slows down the corrosion rate, then the corrosive medium preferentially attacks the highly stressed material, removing the work-hardened layer and aggravating the stress-concentration.

The endurance limit as obtained in laboratory air may be used for the insides of lubricated machinery, structural parts of indoor machinery except in chemically active atmospheres, intrinsically corrosion-resistant materials or cases where we are fully confident of adequate protection. In other cases a substantial safety margin is advisable; usually some guidance can be obtained from previous practice.

See also section 3.7 for static stress corrosion.

3.5 TOUGHNESS AND BRITTLENESS

This concerns the designer because of spectacular failures which have occurred from time to time with little apparent cause, fast-running cracks appearing under steady loading, for instance, in a large molasses tank. It is difficult to imagine anything more steady than a large tank full of viscous liquid. These failures are often at temperatures slightly below 0 °C. In metals, low temperature brittleness seems to be specific to just one form of crystal structure, that of ferritic steels. The problem is how to specify a rational requirement without resorting to unnecessarily costly materials such as austenitic stainless steels, etc., noting that the vast majority of ordinary steel structures survive arctic winters, high stresses and impact loading without trouble.

Toughness is traditionally assessed by hammerblow tests. The current version is the Charpy test in which a small beam, 10 mm square, grooved across the middle of the tension face to leave a net area of 80 mm², is hit by a large hammer at some 2 to 3 m/s. A tough steel will absorb about 80 joules before breaking, that is, 1 J/mm², giving a stringy-looking ductile fracture. At lower temperatures the same material may absorb only 15 to 20 J, which is usually adequate. At temperatures obtained by liquid carbon dioxide or liquid nitrogen (cryogenic temperatures) serious brittleness occurs; the energy absorbed can be as low as 2 to 3 J. What we look for is the transition temperature between ductile and brittle action. Unfortunately this does not seem to be a pure material property; thicker and/or wider specimens and higher hammer speeds show reduced toughness and earlier onset of low-temperature brittleness (see appendix A.4).

An alternative test which is relevant not only to instant fracture but also to

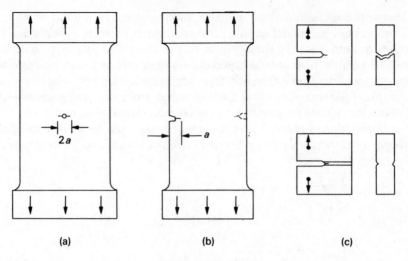

(a)   (b)   (c)

Figure 3.10

progressive failure is the fracture toughness test, using specimens as shown in figure 3.10. This test uses a machined groove merely to locate the notch; the notch is formed by repeated tensile loads so that a short fatigue crack develops. Such a crack is more realistic than a machined groove; it is much sharper and is surrounded by work-hardened material. The crack depth is then measured, with some difficulty in thick plates. Then the spcimen is loaded until it yields to some extent or fractures, or the cyclic loading is continued and the crack growth rate measured.

When loading to instant fracture, Griffith found that the mean stress $\bar{\sigma}$ on the net cross-section was inversely proportional to the square root of the original crack length no matter what this length was. This is very convenient as it is not essential to make the initial crack of a particular length; we can form any reasonable small crack and then measure the depth, making a suitable allowance for the machined starter-notch.

Logically, to measure the fracture strength we should then cool the piece to the lowest actual working temperature and load it until it breaks. If it breaks in a brittle manner the crack depth will be clearly measurable since there is likely to be a visible change between the preparatory fatigue crack and the final break, so we would present the answer as $K_{IC} = \bar{\sigma}\sqrt{a}$ or $\bar{\sigma}\sqrt{(\pi a)}$ (both definitions are in common use).† However, the test is generally done at room temperature, the material is not brittle then, so we only test until we obtain a standard amount of opening, the amount being chosen to indicate the onset of rapid crack-growth.

† The only justification for the two definitions is that the data are usually plotted in log–log form; the presence or absence of $\sqrt{\pi}$ is not unduly noticeable in such plots.

From the measured value and the working stress we should be able to deduce a permissible defect size. Using data from Fuchs and Stephens, [17] p. 300, a component at a steady stress of $\frac{2}{3}\sigma_y$ should be able to stand defects about 3 mm deep. It should however be noted that in a real structure the stresses may develop differently from strain-controlled test conditions, for example, by stress waves [74] feeding energy from adjacent parts into the defect zone. Under fatigue, the calculated defect size is $\approx \frac{1}{4}$ mm.

A connection between K and the actual stresses can be derived using the stress concentration factor for a round-ended slot normal to the mean stress

$$\frac{\sigma_{\text{local}}}{\bar{\sigma}} = 1 + 2 \left(\frac{a}{r}\right)^{1/2} \tag{3.2}$$

Therefore

$$\sigma_{\text{local}} = \bar{\sigma} + 2\frac{K_{\text{IC}}}{\sqrt{(\pi r)}}$$

where r is the root radius of the crack. Since we do not know this radius, we can put in a value equal to the atomic spacing, following Griffith, [7] p. 177. If we do this then the local stress agrees closely with the theoretical tensile strength as derived from surface energy. Alternatively we can take the significant radius as 1 to 2 grain diameters (again following Griffith); then the stress tends to agree with the observed tensile properties. In reference 18 the calculated stress 2 grain diameters below the surface is found to agree well with yield stress. At $\frac{1}{2}$ to 1 grain diameter below the surface it would agree more with the UTS or the final breaking stress. If this correlation proves to be general, fracture toughness will be calculable from tensile properties and grain size.

When it comes to applying toughness data to design we have several problems. One is that many of the data are taken from standard specimens as shown in figure 3.10c. These are much less like a structure than specimens of type a or b; they tend to break with a large shear lip unless they are grooved along the sides as well as at the throat. With plates the problem is unrepresentative aspect ratio; the plate specimens tend to be thin so that crack depth can be measured easily, hence the conditions are not as close to plane strain as in thicker plates. The stresses are only in the ratio of $(3 - 4v)$ to $(3 - v)/(1 + v)$, which ≈ 1.33, but the triaxial effects due to greater width are far more potent than this ratio suggests.

Some steel-works crane hooks are designed on the very logical basis that if the tests do not represent the structure then we alter the structure to fit the test specimens. Accordingly, ladle hooks are laminated from thin plates for which we have test data. As an additional safety precaution the shapes are cut out so that alternate layers run at different grain directions, as in plywood. This of course is costly but is justified by the grave consequences of a failure when

**Ladle
hooks**

**LIFTING GEAR
PRODUCTS
(ENGINEERING) LTD.**

Goliath Works, Petre Street,
Sheffield, S4 8LN

Figure 3.11

carrying many tons of molten steel. Figure 3.11 is reproduced by kind permission of the manufacturers.

The overall size or aspect ratio is also open to debate. Wells [19] quotes
several views that for safety in pipelines the toughness of the material should
be related not only to wall thickness but to the pipe diameter. There is
evidence that even in small pipes cracks will grow much faster than in the
corresponding plate specimens. [20] These points are considered in appendix
A.4.

3.6 SHORT NOTES ON TEMPERATURE PROBLEMS

3.6.1 Sustained High Temperatures; Creep, Oxidation, Diffusion and Migration

Creep of metals is slow yielding under stress, normally in three stages. A
quick initial creep is followed by slow steady creep at a rate predictable from
test data. This merges slowly or almost abruptly into terminal creep leading to

fracture. The time to fracture is also found in data graphs. Because of the complicating effects of oxidation and diffusion it is wise to use materials which not only have sufficient creep resistance but are customary in the particular environment. Data can generally be obtained from makers of suitable materials.

Oxidation, like other forms of corrosion, is accelerated by stress and deflection, notably by varying stresses which break up the protective film.

Migration of elements is important; the commonest is loss of carbon from steel during heat treatment; the carbon migrates to the surface and is oxidised faster than the iron. It might be thought that this leaves the surface material more ductile; unfortunately it is weaker and perhaps under tensile stress due to loss of elements from the crystal lattice. Another common migration is inwards; hydrogen from water or steam affects some materials, causing brittleness. Nitrogen can also cause brittleness at high temperature; in limited amounts it is a valuable hardener (nitriding). An example of internal migration is the production of sigma phase, a concentration of weakening elements at the grain boundaries in stainless steel. Less well known is the effect of zinc on hot stainless steel. [62]

Plastics also have problems, obviously in a much lower temperature range. These include cracking due to loss of volatile plasticiser, creep, effect of ultraviolet light, unsuitable fillers, etc. This is a matter of remembering about temperature and selecting a suitable polymer and filler system. Dark or metallic fillers can be beneficial in reducing light penetration.

3.6.2 Temperature Gradients

These are important inasmuch as they produce stresses and distortions, discussed to some extent in chapters 5 and 6. The distortions can be useful as in bimetal thermostats, slipper-pad bearings, etc. The stresses can be serious; to keep them low we would use material of high thermal conductivity to reduce the gradients and/or of low thermal expansion to reduce the strains for a given temperature difference. It may be useful to remember that a low Young's modulus will tend to give lower stresses. The use of Invar may be appropriate in some cases: this is an iron–nickel alloy of very low thermal expansion in the range −270 °C to +200 °C; above 200 °C it tends to catch up.

Some reduction of stresses can be produced by design, placing cooling passages in accordance with the rate of heat input; for instance in turbine blades the heat input is highest near the leading edge; it may help to put cooling ducts near the leading edge and deliberately omit cooling passages in the centre. In steam turbine casings it helps to make cuts in the outer edge so that during warm-up the inner surface can expand without bowing the whole side inwards, see figure 3.12, adapted from reference 21.

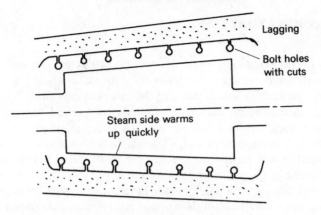

Figure 3.12

3.6.3 Cyclic Temperatures

Under cyclic temperatures we get just about all the disadvantages. Temperature gradients during rise and fall give reversing stresses, the high temperatures may remove beneficial work-hardening at the base of incipient
cracks, the expansion and contraction may flake off any protective oxide
layers. Some of the precautions a designer can take will be discussed under
pressure vessels. One way in which the designer can avoid frequent fluctuations is to insist on proportional heating control rather than on–off; another is
to try to ensure gradual start-up and shut-down procedures.

A particular problem occurs sometimes with austenitic stainless steels. The
reader is presumably aware of the need for certain stabilising elements added
to the 18/8 type of stainless steel in order to produce grades intended to be
welded. Even so some degree of transformation is likely, which suggests that
the austenitic form is not entirely stable. The evidence for this is scattered; it is
known that the magnetic permeability increases due to welding and to a lesser
extent due to cold-working. Also stainless steels tend to rust if cut with
abrasive high-speed cut-off wheels.

The problem which the author has met on two separate occasions is substantial distortion when a large temperature change takes place repeatedly
under high stress. This occurred, in one case at very low temperatures, in
another at above 120 °C. It is thought that when partial transformation occurs
under stress, freedom of movement is exceptionally high. Furthermore it is
possible that tensile stress encourages one transformation (the one which
causes dilatation) while compressive stress encourages the reverse one, in
accordance with Le Chatelier's principle. In bending, distortion could occur
twice as fast as in compression or tension alone since there would be an
opportunity both on the rise and on the fall of temperature.

3.7 CORROSION AND STRESS

The designer is expected to be aware of corrosive problems. The main choices in dealing with corrosion are intrinsically resistant materials, surface protection, cathodic protection and minimising the extent of exposure. A relatively new concept is that some materials normally considered adequate for a certain environment become corrodible when under stress above a certain value, the stress corrosion threshold. This value is remarkably low. Corrosion is further aided by differences of stress that inevitably occur in structures, the most highly stressed parts corroding preferentially. Most corrosion problems are well known; the important point made here is the threshold problem. This could crop up when a well-established design is taken to higher stresses thanks to better methods of calculation, then the corrosion problem could turn up unexpectedly. Apart from specific chemicals for which suitable materials must be used, the most severe general atmospheres are airborne sea-water splash and spray and sulphurous industrial fumes. Buried work is liable to suffer due to some aggressive ground-water conditions. The designer cannot always stick to the very low stresses (20 to 30 N/mm^2) used in the past; fortunately modern protective coatings, cathodic protection, etc., can deal with many of the troubles. We need to look out for specific sensitivities, for example, sulphur or chlorine on some stainless steels, concentrated acid or salts in tank bottoms, loss of inhibitors by age and heat, water in oil tanks.

We conclude this section with a few design points. Corrosion due to moisture is likely to defeat protection in the long run; good design can prolong life. In figure 3.13a the bent member riveted or spot-welded to the next part has crevices liable to hold moisture and start corrosion. Corner A is more likely to be under stress than corner B so if A can be put into a dry zone this is to be preferred. In outdoor structures there is the obvious water trap which is likely to destroy the paint-film prematurely (b). This should either have generous drain holes or be replaced by a channel. The old-fashioned I-joist with tapered flanges may sometimes be better than the modern flat-flanged universal beam (c and d) from the drainage point of view. Access for painting

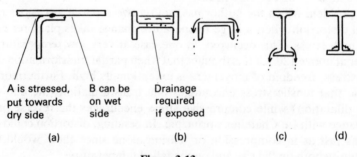

A is stressed,   B can be   Drainage
put towards   on wet   required
dry side   side   if exposed

(a)   (b)   (c)   (d)

Figure 3.13

and inspection is worth bearing in mind. The use of dissimilar materials invites galvanic corrosion if the two form a short-circuited battery; it may be wise to go to considerable trouble to insulate the two materials electrically from each other if their use cannot be avoided.

Example 3.1

A material has a UTS of 320 N/mm², yield point 260 N/mm², endurance limit in rotating bending of 150 N/mm². Replot figure 3.14 and find

(a) the endurance limit in repeated tension;
(b) the endurance limit in unidirectional repeated torsion;
(c) the fully alternating stress to cause probable failure in 10⁵ cycles;
(d) the probable life under a stress cycle ranging from 120 N/mm² compression to 180 N/mm² tension.

Working on figure 3.14, answer (a) = 204 N/mm², answer (b) = 204/√3 = 118 N/mm² (see section 3.4.4), answer (c) = ± 180 N/mm².
For answer (d) we show the stress cycle on the Smith diagram, using the mean value of 30 (midway between 180 and −120). The argument is that this is equivalent to a fully reversed stress (d) obtained by drawing from the apex to the axis as shown; this stress would cause failure at 10⁵·⁵ cycles, that is, ≈ 300 000 cycles. Don't use this idea below 5000 cycles, the data are uncertain.

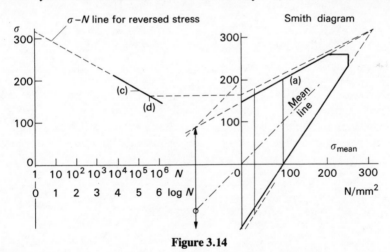

Figure 3.14

Problem 3.1

Using the same diagrams, what life should we expect from a stress cycle of ± 220 N/mm² (tension/compression), +220 to −100 N/mm² tension, +220 to 0 N/mm² tension?

Ans. 4000, 25 000, 250 000 cycles.

Comment: by extending the Smith diagram into the compressive region we can attempt to find how much compressive stress is needed to lengthen fatigue life. For instance, a compressive stress of 110 N/mm^2 has been put on the diagram; this suggests that we can then impose a stress of up to $\pm$ 200 N/mm^2 and still obtain infinite life.

4 STRESS CONCENTRATIONS

Stress concentrations due to changes of shape arise in most designs. An awareness of the severity of such concentrations is helpful in making design decisions. It is unrealistic to say that all corners must be avoided. We must accept that there will be stress-raisers, for good reasons.

The main stress-raising features are steps, grooves and holes. Their severity depends chiefly on the radius, in other words the sharpness, and secondarily on the relative change of cross-section normal to the lines of force. It is helpful to think of components in terms of lines of force as in magnetism. Where the lines curve they produce a transverse force since a tensile line of force tries to become straight and vice versa. Where the lines crowd together we find high stresses. The local stress is calculated by taking the nominal stress at the net cross-section and multiplying it by a stress concentration factor, except in a few cases such as keyways in shafts where it is usual to use the complete shaft diameter as basis and let the stress concnetration factor include both shape effect and diminished cross-section.

SCF values are determined in some cases by calculation, sometimes by photoelastic models, sometimes by electrical analogues. When SCF values are calculated from endurance tests they are called strength reduction factors. These are of limited value since they may be affected by size, speed, prior load and various errors, see appendices A.1 to 5. For many practical cases the data do not fit. The form of presentation adopted here is intended to help in interpolating when a design falls between the available solutions; the main ratios are corner radius to base diameter and to step height.

Data from different sources show discrepancies. It is not always possible to give the reason; sometimes photoelastic data, on examination, are found to have been deduced from photographs with very few fringes, requiring extrapolation by means of the optical density. At other times the material may have been strained beyond the linear range in order to give many fringes. Calculations and electrical methods have limited resolution in sharply curved regions. Luckily, SCF values rarely disagree by more than 10 per cent with a few notable exceptions. One set of values for an infinite perforated sheet is suspect, giving SCF values much lower than the sheet cut up into strips as in figure 4.12. Another set is more than suspect; for a certain pattern of holes it claims that an SCF of zero is possible! The reader is warned that there is no guarantee that published data are safe.

4.1 SHOULDERED SHAFTS (STEPPED SHAFTS WITH SHOULDER FILLETS)

The stress concentration factors for this common detail in bending and torsion are shown in figure 4.1. The form of presentation adopted brings out the primary importance of the root radius. Torsion data are somewhat limited; in cases of need an estimate can be obtained by the rule

$$\text{SCF in torsion} \approx \frac{1}{2} + \frac{1}{2} \times \text{bending SCF} \tag{4.1}$$

Data for tension are not often needed; the SCF in tension is consistently between 4 and 10 per cent greater than that in bending. The data marked (P) are from Peterson, [22] the others by Allison. [23]

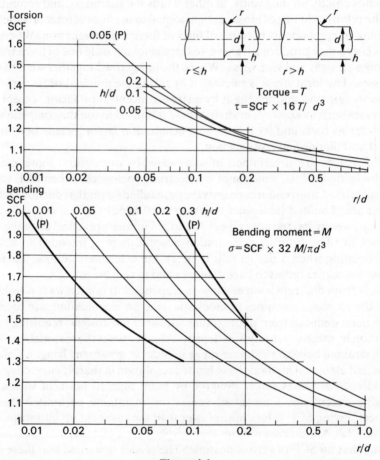

Figure 4.1

For h/d from 0.01 to 0.1 the curves fit equation 4.2

$$\text{SCF in bending} \approx 1 + 1.2 \left(\frac{d}{r}\right)^{1/2}\left(\frac{h}{d} + 0.07\right) \tag{4.2}$$

For larger steps, that is, $h/d > 0.2$ but $r/d < 1$

$$\text{SCF} \approx 0.725 + 0.54 \left(\frac{d}{r}\right)^{1/2} \tag{4.3}$$

These equations reflect the physical aspects of the sharpness of the corner and how a small step height reduces the fanning-out and hence the severity.

The graphs and equations imply that in a sharp corner the stress is infinite. This is obviously impractical inasmuch as many sharp-cornered components survive without failing, at quite high loadings. In practice corners are rarely sharper than 0.1 mm radius. Furthermore, the grain size of the material reduces the effect of a corner. Long before fracture mechanics became popular this effect was spoken of in terms of notch sensitivity, hard materials of fine grain being more sensitive than softer ones. The situation is further complicated by local yielding which modifies the stresses and also modifies the actual shape, sometimes quite significantly.

A series of geometrically similar shouldered shaft specimens were used in what is thought to be the first demonstration of size effect in fatigue. A number of samples in three different sizes were made from the same material and tested in rotating bending to establish the endurance limit. This was compared with the endurance limit for smooth specimens of negligible stress concentration, that is, SCF = 1. The ratio of the smooth endurance limit stress to the stress in the plain part of the shaft for specimens just surviving a very large number of cycles is described as the strength reduction factor or apparent SCF. Figure 4.2 shows that the smallest specimen stood stresses almost equal to a smooth piece; small size offsets stress concentration; a very important matter when we use test data from small specimens. [24]

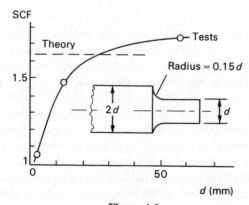

Figure 4.2

It is not essential for the shouldered shaft to be made in one piece; if two components are firmly press-fitted or shrunk together and subjected to rotating bending they will either rub with likelihood of fretting which rubs away the oxide layer and can start a corrosive crack or, if the fit is tight enough not to slip, then the assembly behaves as one shouldered component. This was first shown by Kuhnel [25] in connection with railway axles. The material had a UTS of 500 to 600 N/mm^2, its endurance limit being $\pm$ 230 N/mm^2. The specimens consisted of simulated axles with pressed-on hubs, figure 4.3a, b and c. Specimens a and b gave results agreeing with the endurance limit. Specimen type c, repeated several times, gave endurance stresses of only $\pm$130 to $\pm$150 N/mm^2. Then the experiments were repeated with solid shafts turned to a shape equal to the press-fitted assemblies and gave very similar results. The conclusion which is not unduly surprising is that as far as corner stresses are concerned the assembled set should be treated as a solid component. This conclusion has been amply verified by much additional work, see Peterson and Wahl [26]. If it had been taken to heart and applied 20 years later, the failure discussed in chapter 10, section 10.2, would not have happened.

If we need the effect of a sharp corner to suit the purpose of the machine, we can produce a corner radius by undercutting the face, the diameter or both. The SCF will be somewhat greater than for the right-angled shoulder fillet because the lines of force have to turn through a larger angle. Figure 4.4 shows a series of arrangements of equal r/d and h/d ratios in order of increasing SCF. The first arrangement should have the lowest SCF though data are lacking. For comparison, a long groove and a short U groove are also shown. Forms d and e are not fully comparable with the others on a diameter basis but are shown because they are common practical details for manufacturing reasons.

4.2 GROOVED SHAFTS

Practical grooves may be rectangular with slightly rounded corners, for example, circlip grooves; rather more rounded corners may be found in O ring grooves. Fortunately it is very rarely necessary to have these in regions of high bending moment or high torque. Deep grooves of U shape or with flat base and corner radii are rare in solid parts as they are difficult to make but they feature regularly in shrink-fitted assemblies of gearwheels, steam turbine wheels, etc. As shown in section 4.1, the shrunk-on parts make the assembly behave as a solid body, the corners becoming relatively severe. Figure 4.5 shows SCF values for torsion and bending of shafts with relatively sharp grooves, that is, small r/d values, replotted from Peterson. [22] The bending values for less severe cases are given in figure 4.6, with an extended vertical scale. Any torsional values not readily available can be estimated from equation 4.1, that is, $\frac{1}{2} + \frac{1}{2} \times$ bending SCF.

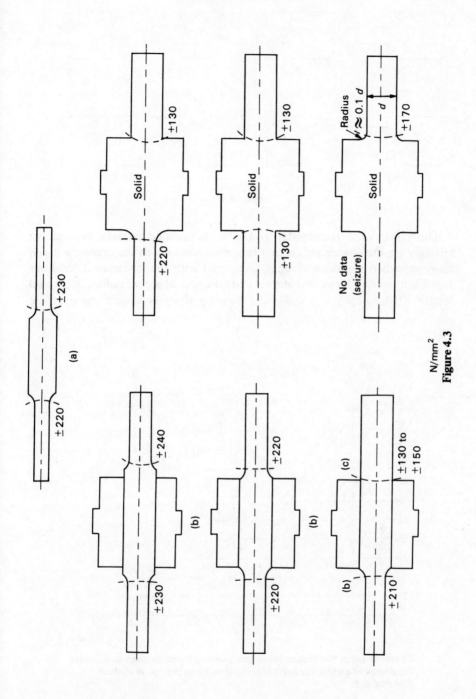

N/mm²
Figure 4.3

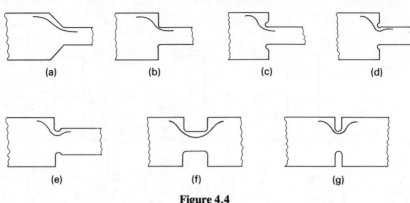

Figure 4.4

The effect of a rectangular groove with rounded corners depends so strongly on the corner radius that this must continue as the common basis; however when a groove is small compared with the diameter it becomes feasible to ignore the shaft diameter and take note of corner radius, depth and length. If the depth $h < 0.03d$ the bending stress is so uniform over the

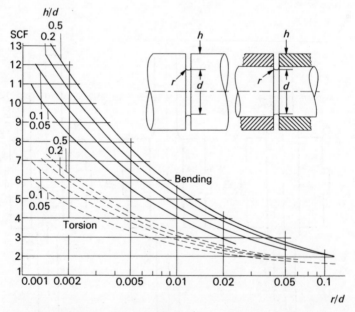

Stress concentration factors for deeply grooved shafts (U-shaped grooves)
Solid lines — bending; dashed lines — torsion. Basis , stress in shaft of
diameter d.

Figure 4.5

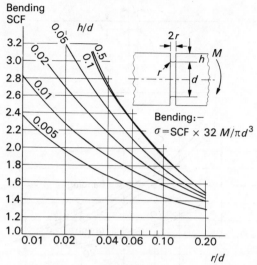

Figure 4.6

affected region that the SCF can be taken as valid for tensile load as well as for bending load. Figure 4.7 gives a range of SCF values for grooves of length ≥ ¼d. Note that these are very close to the values for a shouldered shaft. For small grooves figure 4.8 gives a fuller story. Over the range of data shown there

$$\text{Bending SCF} = 2.4 \left(\frac{r}{h}\right)^{-0.45} \left(\frac{r}{L}\right)^{0.083} \tag{4.4}$$

Torsion SCF again ≈ ½ + ½ × bending SCF

These data are replotted from ESDU items 69020, 69021, 79032. [27]

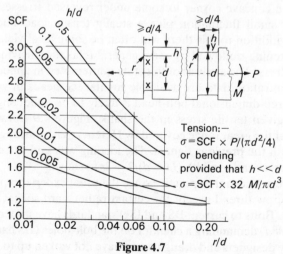

Figure 4.7

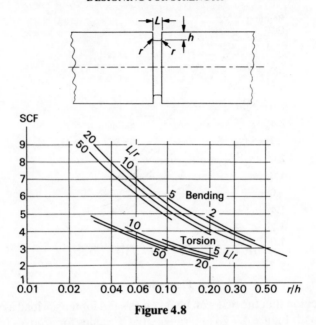

Figure 4.8

4.3 BOLT-HEADS

Bolt-heads show a much higher stress concentration factor than the geometrically similar shouldered shafts since the lines of force have to turn
though 180° but this is offset by the loading conditions. It is virtually im-
possible for the concave corner to come under reversed stresses; the usual
condition is a small fluctuation plus a steady tensile load due to initial
tightening. In addition to this there will often be some local yielding during
the initial tightening, followed by some general relaxation leaving a favourable residual stress. The analysis of this state is attempted in appendix A.5.

Stress concentration data are available for flat tee-heads in various proportions. A three-dimensional bolt-head is likely to be under slightly lower
stresses for a given tensile stress in the shank. Figure 4.9 shows the most
practical part of the great range covered by Hetenyi, [28] rearranged for easy
comparison with the previous figures. The loading was said to be uniformly
distributed over a band of width $(D - d)/2 - 2r$ as shown.

This is not very relevant in practice since the usual experience is that bolts
break at the screw thread except for certain rather hard socket cap screws
(Allen screws). Bolts to current BS and ISO standards have underhead fillets
up to about $0.08d$, demanding a chamfer on the bolt-holes (BS handbook 18,
p.5/105); many designers and draughtsmen have not woken up to this yet.

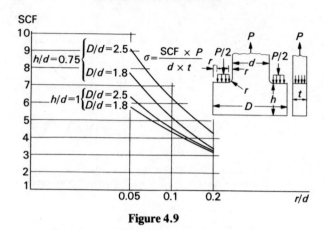

Figure 4.9

4.4 SCREW THREADS

The working conditions of screw threads vary over an appreciable range of shapes and loading cycles. It is rare for any given assembly to be under reversed loading; even if it is, the lines of force follow different paths in push and in pull, figure 4.10a. The configurations found in practice include the plain stud, various forms of run-out groove or thread termination to suit the form of production by die-head, rolling, etc.

The stress concentration at the thread root radius is usually aggravated by the tendency of the majority of the load to be taken by the first two turns of thread. To explain this, compare the two arrangements in figure 4.11. The bottle-screw or turnbuckle has both the rod and the nut in tension. Under load the screw and nut both get longer; thanks to the gradual lead-in the extensions will be similar and the load will transfer progressively, shared over many turns of thread. In the stirrup arrangement the opposite takes place. The screw lengthens under the load but the nut is in compression and gets shorter. With a little imagination we can see that the force will pass predominantly through the first contacting flank. Various forms of relieved nut have been proposed but these are mostly inconvenient as they occupy more space. The most successful method is that used on steam turbine casings. The nut is tapered so that there is a slight clearance at the critical end. As the nut is tightened, the elastic stretch of the bolt and the shortening of the nut take up the clearance, giving uniform load transfer if the taper has been calculated and made correctly. The taper required is quite small; it may occur spontaneously during hand-tapping.

In view of the many variables discussed above it is not surprising that comprehensive data are lacking. The most useful selection of which the author is aware is ESDU item 68045 [29] which surveys test data in fluctuating

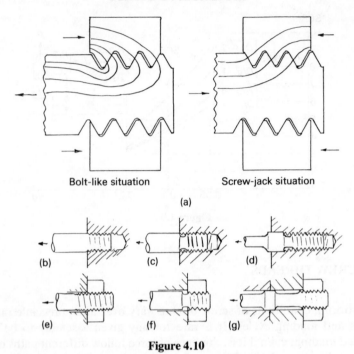

Bolt-like situation　　　　　Screw-jack situation

(a)

(b)　　　　　(c)　　　　　(d)

(e)　　　　　(f)　　　　　(g)

Figure 4.10

tension, the tests being arranged as in figure 4.10e. A number of different materials ranging in UTS from 550 to 1700 N/mm² at nominal mean stresses between 0.25 and 0.4 of UTS gave a fairly consistent picture. The stress range for 10^6 cycles life depended mainly on the UTS and on the method of manufacture, any dependence on mean stress being hidden by the general scatter. Expressing the fluctuation as the amplitude of nominal stress (load/ thread root area) gave an endurance limit of ± 0.2 to 0.22 of UTS for rolled threads and the remarkable result that if these were heat-treated after rolling the endurance limit fell to ± 0.05 of UTS, possibly due to decarburisation and/or loss of favourable residual stress. Machined threads gave an endurance limit of ± 0.1 of UTS.

It is important to remember that the thread run-out (type e, figure 4.10) is favourable to long fatigue life; type b would show greater stresses, shorter life. This series applied to standard V form threads with rounded roots. There is little difference between the current metric and inch thread forms in this respect. Very rounded threads and buttress threads may be expected to be slightly stronger in fatigue, other things being equal.

It is possible to derive an effective stress concentration factor from these figures (strength reduction factor), by the following argument.

For a mean stress of 0.3 of UTS in a smooth component we would expect a permissible stress range from say 0.2 to 0.7 of UTS, in other words an

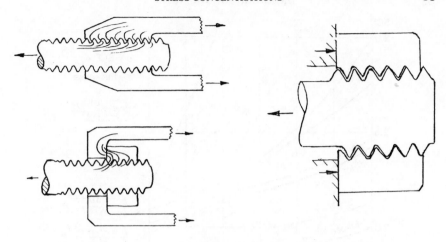

Figure 4.11

amplitude of $\pm$ 0.25 UTS (by Goodman or Smith diagram) hence the strength reduction factor or effective SCF $\approx$ 2.5 for cut threads, 1.25 for rolled threads provided they are not given an unfavourable heat treatment. Very hard high-tensile bolts should be treated with caution; a value of SCF = 4 is suggested (as in references 49 and 50 for all screw threads).

For tapped holes there is little specific information. The designer may note figure 4.10d, an arrangement recommended for steel studs in aluminium alloys using a thread engagement 3 to 4 bolt diameters long, with a counter-bore to discourage shearing of the first thread turn and raising of the surface.

The design of bolted assemblies is discussed in chapter 6. It will be shown that it is usually possible to stress bolts very highly because in an assembly the load in the bolt is fairly steady, load variations being largely absorbed by transfer of forces within the structure.

4.5 HOLES IN BARS

Figure 4.12 shows collected data from various sources on the SCF of bars with holes of various shapes and spacings. The basis is the stress at the net cross-section, that is, maximum stress = SCF $\times$ $P/(b-d)t$.

A single hole in a rectangular bar has SCF values ranging from 3 downwards, shown on the second curve of figure 4.12. These values $\approx$ 3 $-$ $1\frac{1}{4}(d/b)^{0.6}$.

Attention is drawn to the benefits gained by holes in a row. If we have a

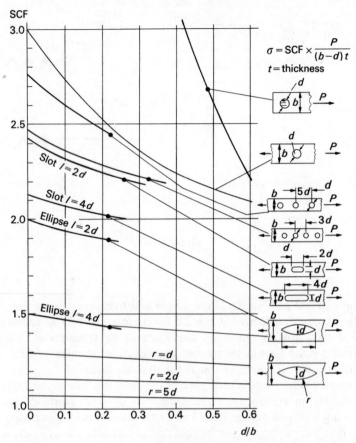

Figure 4.12

functional hole (not load-bearing however) we can reduce the SCF by drilling holes in line with it, preferably of smaller size. A suitable series of holes can reduce the SCF to values similar to the 4:1 ellipse.

The two penultimate shapes are ellipses; the last three forms show the benefits due to a gradual widening. The data for this were derived from the converse, a strip with radiused flanks. The last five shapes are not easily machined by standard cutters but can be produced readily by casting, punching, electrical machining (either spark or electrolytic), etc.

A slot across the load line has two effects, one due to the width and one due to the sharpness of the corners (figure 4.13). From the data in ESDU item 69020 the author has derived the expression

$$\text{SCF} \approx \left(2.6 - \frac{2w}{b}\right)\left(\frac{w}{r}\right)^{1/3} \qquad \text{provided } w < 0.8b \qquad (4.5)$$

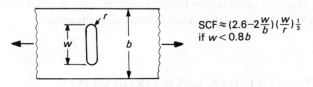

$$\text{SCF} \approx (2.6 - 2\tfrac{w}{b})(\tfrac{w}{r})\tfrac{1}{3}$$
if $w < 0.8b$

Figure 4.13

In biaxial tension the SCF is always smaller than in the corresponding uniaxial case; a single hole in a much larger member with equal tension in all directions has an SCF = 2. On the other hand a hole in a shear panel has an SCF = 4. This should not surprise us if we remember that a pure shear state has tension and compression at 45° to the shear axes so there is a double tendency to ovalise the hole.

If a load is applied via a hole in a tension member we call it a lug. Figure 4.14 shows how the lines of force tend to concentrate firstly at the contact face, secondly at the minimum section, provided that a sensible amount of material has been provided behind the hole. This rear distance should generally be over $1\tfrac{1}{4}d$; in heavy steelwork where the edge is rough and corrosion is possible, edge distances of $2d$ or more are recommended, especially in the case of riveted joints where the swelling of the rivets adds extra stress.

Some references quote stress concentration factors of 10 or 20; this applies to cases where fretting can occur between an aluminium alloy lug and a steel pin. In general machine design with lubricated (or at least corrosion-protected) joints and well-fitting (slide-fit) pins, the highest stress depends on the local bearing stress usually expressed as P/dt and the crowding effect at the net section. The author has reanalysed the published data on this basis. The highest stress comes at about 45° and amounts to

$$\sigma_{\max} = \frac{P}{dt} + \frac{1.6P}{(b-d)t} \tag{4.6}$$

When expressed as an SCF based on the net section, SCF = $b/d + 0.6$.

If the pin is a loose fit, the local compressive stress rises but the highest tensile stress increases only slightly. Where fretting danger is expected, some possible precautions are relieved sides, press-fitted bushes of stronger material (but look out for intermetallic corrosion), or certain surface treatments (of limited life).

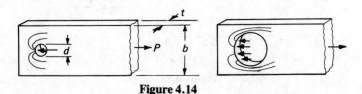

Figure 4.14

Holes in flat bars are only the limiting case of holes in other members; the SCF values shown above are a guide to most such situations.

4.6 SHAFTS WITH HOLES, KEYWAYS OR SPLINES

Oilways, keyways and splines in shafts are necessary features and sometimes must occur at points of high stress in torsion or bending. If a shaft is in pure tension and has a transverse hole, the SCF is similar to the equivalent flat bar case. If preferred it can be taken as about 10 per cent greater than the bending case quoted below.

Because the calculation of the bending and torsional properties of a shaft with a transverse hole or keyway is very inconvenient, the values given are not the local SCF based on the net section but the total effect of reduced cross-section properties plus local stress concentration; the maximum stress is SCF times the stress in a plain shaft with no hole, etc. Figure 4.15 shows the torsional and bending SCF for solid and hollow shafts. [30]

Shafts with keyways are a complex subject. It has been shown that in tightly fitted assemblies a substantial part of the torque is transmitted by friction resulting from the elastic distortion of the shaft and hub assembly. If the key is of standard design and is of ample length or where that is difficult if two keys

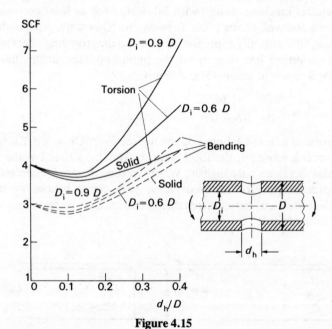

Figure 4.15

at 90° positions are used, the critical region is likely to be at the end of the keyway. The three basic forms are the face-milled or sled runner type S, figure 4.16, the end-milled or bathtub form B and the Woodruff type W, for light duties. A new proposal published recently [31] combines the features of S and B, cut by using an end mill with progressive run-out, shown at SB. The B and W forms have the advantage of securely trapping the key.

Some older books suggest that SCF values lie around 1.6; this is thought to be based on small-scale tests subject to size effect. For torsion an SCF value of 3 is recommended for design purposes, for types S and B. In bending, type B seems to have an SCF of around 3 again, provided the keyway end is well away from a shaft shoulder. If the keyway comes right up to a shoulder of typical height or worse still if it is milled into the shoulder then Fessler's data [32] indicate that SCF $\approx$ 4.5 to 5.

Splined shafts come in two forms, X and Y, figure 4.16. The X form is similar to keyways of the S type. The Y form has a very gradual transition and should have lower SCF values. Furthermore many splines are made by form-rolling which leaves a highly favourable residual stress. For machined splines of form X an SCF = 3 seems advisable for design purposes, especially if high-tensile notch-sensitive material is being used.

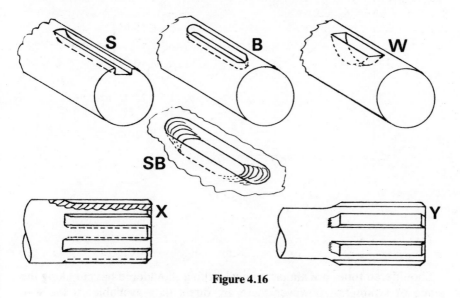

Figure 4.16

4.7 FLAT BARS WITH ENLARGED PORTIONS; SHOULDERED TUBES

This type of detail is found in castings such as baseplates, rocking levers, etc., also in many forged or machined components generally. When the enlarge-

ment is long, two or more times the thickness t, figure 4.17, then the SCF
is similar to that for shafts but slightly higher. When the enlarged portion
is short, the stress is reduced because the lines of force diverge less. The
published data vary slightly more than usual, those in Peterson [22] running
up to 10 per cent higher or lower than ESDU item 69020. Most of the data stop
at $r/t = 0.2$; since in castings and forgings r/t would generally be 0.3 to 0.5, the
data have been extrapolated using the author's treatment; these data are
distinguished by dashed lines. The values in figure 4.17 are a compromise;
they fall within ± 5 per cent of the highest published data.

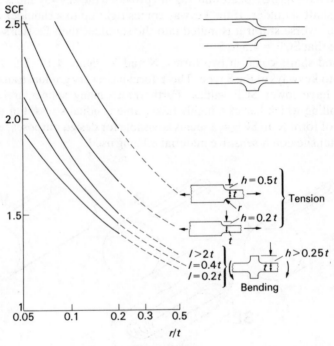

Figure 4.17

Shouldered tubes are almost equivalent to a shouldered bar cut along the
plane of symmetry; however there are direct data available. If the wall
thickness is much smaller than the diameter then the logical presentation will
refer to the wall thickness and step height. The SCF values for tension and
bending are then very similar anyway but in torsion the SCF is considerably
lower. The basis is as usual the stress in the thinner member alone under the
appropriate loading. Figure 4.18 shows tension values rearranged from
ESDU item 69021, also torsion values from reference 33.

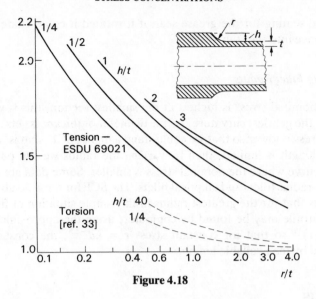

Figure 4.18

4.8 SOME STRESS-REDUCING DESIGN FEATURES

(a) *Protective Groove*

If a shaft needs a shoulder for location purposes, etc., and the root radius is restricted by the adjacent member then the SCF can be reduced as in section 4.7 by shortening the enlarged shoulder into a mere collar, figure 4.19a. One would make $r_2 > r_1$ because of the larger angle of deviation.

(b) *Spacer Piece*

An alternative is to make the fillet radius as generous as space allows and provide location by a loose piece, figure 4.19b. This method is sometimes used in automotive hubs. It is advisable to secure the loose piece firmly if it is

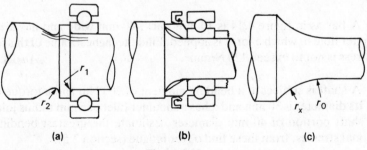

Figure 4.19

used as the seating for the grease seal; if it rotated it could chafe a stress-raising groove into the stub-shaft.

(c) *Varying Fillet Radius*

Since the nominal stress is highest at the smallest section, this is where we should put the gentlest curvature. As the diameter or thickness gets larger the nominal stress is lower so that we could stand a higher SCF, that is, a sharper radius. If length is limited, then by varying the radius we can put a more generous curve where the nominal stress is smaller. Some data are available for double-radius fillets and elliptical fillets. The SCF for these is substantially the same as that for the greatest radius in the simple situation of figure 4.1. The best profile may be found by computer, using the approximation that SCF $\propto (d/r)^{1/2}$ so that for constant stress $r = kd^{-2.5}$, the constant being determined by the available length.

PROBLEMS

4.1 A shaft of 400 mm outside diameter has grooves 20 mm deep, 10 mm wide, 1 mm root radius. Find the SCF in bending and torsion from figure 4.8.

Ans. 7.15, 4.0

4.2 The same size of shaft has a groove of root radius 10 mm, width and depth 20 mm. Find the SCF in bending and torsion, from figure 4.5.

Ans. 2.6, 2.0

4.3 The shaft of problem 4.2 has shrunk-on discs added to it resulting in an outside diameter of 720 mm. Find the new SCF in bending and torsion.

Ans. 3.6, 2.35

4.4 If the shaft of problem 4.3 is altered so that it has a U groove of 1 mm root radius, base diameter 398 mm, find the new SCF in bending and torsion.

Ans. 10.4, 5.7

4.5 A bar as in figure 4.14 is 10 mm thick, 40 mm wide and has a 12 mm diameter hole to which a force is applied. Find the highest value of this force if the stress is not to exceed 120 N/mm^2.   *Ans.* 8.54 kN

4.6 A shaft is subject to a bending moment of 200 Nm and a torque of 300 Nm. Its diameter is 30 mm and it has a shoulder fillet of 6 mm radius joining it to a shaft portion of 40 mm diameter. Estimate the greatest bending and torsional stresses, from these find σ_e for fatigue (section 2.5).

Ans. 109, 67, 160 N/mm^2

5 STRUCTURES IN TENSION AND COMPRESSION — FRAME STRUCTURES

5.1 SIMPLE WALL BRACKETS

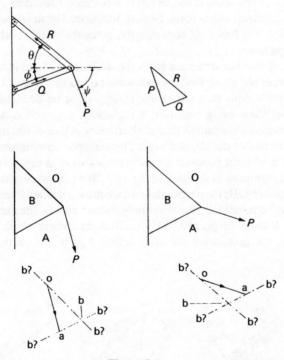

Figure 5.1

Figure 5.1 shows a simple wall bracket carrying a load in the plane of the paper. The forces in the two members are obtainable by resolving vertically and horizontally. If ϕ were zero and ψ 90°, the case could be solved by inspection. In the general case shown we may prefer the graphical approach, namely the triangle of forces. A convenient routine for finding this is to use Bow's notation; this will be found particularly helpful in building up the force diagram for an elaborate structure. Here it is shown twice, for the same structure but different load directions. We label the spaces between the loads and members; the outside is often called O. Proceeding clockwise round the junction we cross from O to A via the load P, so we draw the load vector **oa**, equal to P to a convenient scale. Crossing from A to B we know the direction of the force in AB but not its sign or magnitude, so we draw a general line parallel to AB, calling it ab?. Similarly, we draw b?o parallel to BO. Where these cross we have point b and we can read off forces **ab** and **bo** in magnitude and direction.

The vector **ab** shows the direction of the force being exerted on the junction by the rod AB thus showing that AB is in compression; similarly for the other cases; in the second picture AB is in tension. In most structures this can be confirmed from common sense but in complicated structures the convention gives the true answer of the force on each junction. The anticlockwise convention, provided it is followed consistently, gives the same result as the clockwise one used here.

A similar vector construction gives the deflection of the load point OAB. From the forces **ob**, **ab** we find the extensions or contractions of the members. This of course means that the cross-sections must be decided first, at least provisionally; then using equation 2.2 gives us Δ_{OB} and Δ_{AB}. Figure 5.2a shows that merely extending OB and shortening AB does not make sense; we have to make use of the dashed arcs. These displacements are grossly exaggerated; we could not produce an accurate solution in this way. We need to use the displacements as vectors. In figure 5.2b we draw the extension Δ_{OB} in line with member OB; then from the end we draw a dashed line at right angles to represent the rotation. This is actually a small part of the arc shown in (a). This line is a locus of point OAB. To find just where on this locus OAB actually sits, we now draw the contraction Δ_{AB} in line with member AB

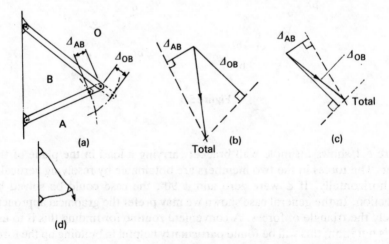

Figure 5.2

together with a dashed line at right angles. This also is a locus of OAB. Where these lines cross is the actual point OAB; the total displacement is the vector from the starting point to this final point. In (c) we show how this construction works out in the second case of figure 5.1. The cross-sections assumed for these examples were roughly based on equal stresses. With rigid joints, the bracket would behave as at (d); the forces would be similar but modified by the bending effect.

5.2 MORE GENERAL FRAMES

This section shows how we can find the forces in a more complex structure provided it is free from redundancy. A redundant structure in this sense has more than the minimum number of members; the forces in such a structure depend on the relative stiffness of the members, see section 5.5

Figure 5.3 shows a simple girder or truss with a central load, on frictionless supports. We find the reactions and label the spaces. Starting at O and going round junction OAC clockwise

$$\text{vector } \mathbf{oa} = \frac{P}{2} \text{ upwards}$$

$$\text{vector } \mathbf{ac} \text{ is bound to be horizontal}$$

$$\text{vector } \mathbf{co} \text{ is parallel to member CO}$$

The meeting point of these lines gives point c.
A similar process round junction OBD gives points b and d, completing the diagram. The diagram tells us that ab = P which we knew already by inspection. It also tells us that cd = P and that CD is in compression, by studying point CDO. The *downward* vector **cd** shows that member CD pushes *down* on to junction CDO.

If there is friction at the supports then we must find the direction of the frictional action on the structure. We expect AC and BD to be in compression and get shorter. Friction will partly oppose this, so it will act outwards. Figure 5.4 shows the diagram with friction included. If there is a horizontal force $\mu P/2$ at one side then by horizontal equilibrium there is an equal and opposite force at the other support; the friction force has no effect on the vertical forces.

Bow's notation has great diagnostic powers. As a first example consider a mast 40 m high, hinged at the base and secured by three ropes equally spaced, fastened down 20 m away from the base. The wind force = P at the top, in any direction. We require the maximum rope force in terms of P, assuming the down-wind rope to go slack.

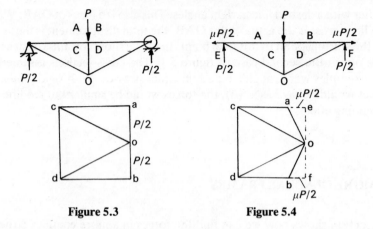

Figure 5.3   Figure 5.4

In figure 5.5 we start with the plan view of the three ropes and the force P at some angle θ. We *sketch* the force diagrams for $\theta = 30$, 60 and 90°; then we superimpose the diagrams. The composite diagram clearly shows that the greatest possible oa value $= P \sec 30°$ and occurs when $\theta = 30°$. From a casual inspection we might have guessed that the greatest horizontal force would be equal to P, when $\theta = 60°$ or $0°$. A side view then shows how the rope force can be found from the horizontal force.

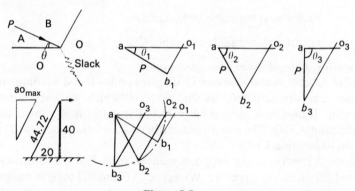

Figure 5.5

The next example is left to the reader. Draw the hydraulic loader mechanism in the three positions shown. For the middle position find the forces in all the members in terms of P. Then draw the force diagrams for the other positions and note whether any of the forces are greater than in the mid-position (figure 5.6). Reactions at the fixings are not required.

Ans. ob $= 3.42P$, oc $= 2.9P$, ab $= 3.27P$, bc $= 0.9P$;
top position oc $= 3.05P$, lower position bc $= 1.3P$.

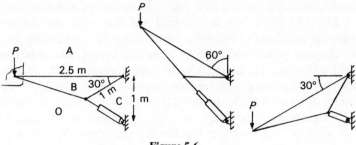

Figure 5.6

Dealing with multiple loads is no problem; the more complicated the set-up the more we appreciate the elegance of the graphical approach, whether drawn accurately or sketched and solved by calculation thereafter. The roof truss in figure 5.7a will be solved, then we shall show how to cope with added complications. When the load is symmetrical we need not find the reactions in advance. We start by setting out the load vectors **oa**, **ab**, **bc**. The reaction **cd** is simply half-way back to o. The other points follow in the usual way. Points e and h coincide; this often happens in symmetrical structures. If *adjacent* points coincide this shows that we have a load-free member between the spaces concerned.

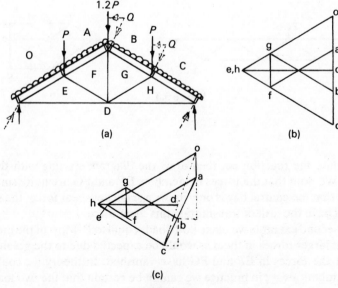

Figure 5.7

To show how easily we deal with angled forces, let us add wind loads Q to AB and BC. The forces are now angled and increased; the reactions are altered too. It is helpful to calculate the reactions first, by taking moments. The new diagram is shown in figure 5.7c. **oa** is as before, **ab** and **bc** have new values and directions. The reactions point back towards o. If the friction is not shared out at the same angle, the reaction line will have a small kink; it will not make much difference to the forces. The student is advised to attempt the diagram independently and then compare with the solution in figure 5.7c.

The powers of this method are further illustrated by subjecting the structure shown in figure 5.8 to several different loadings. In the first example

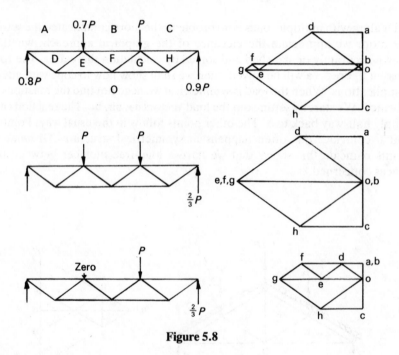

Figure 5.8

we calculate the reaction **co**, then draw the diagram starting with the load vectors. We note that the forces in members EF and FG are quite small; the reason is that the central bay is only under a small net shear force; the sloping members have the task of transmitting this shear force.

In the second example we make both loads equal to P. Most of the members now have larger forces in them as would be expected due to the greater total load, but the forces in EF and FG have vanished. In theory we could take these members away; in practice we cannot be certain that the two loads will always be equal, nor can we be certain that the members AD, BF and CH will always be dead in line. The structure would be metastable; any slight deviation would cause it to fold up at the joints.

The third example shows the forces with one of the loads removed altogether; now the total load is less than ever but members EF, FG have the highest loads found so far. This shows the need to analyse a structure under all possible loading cases, not just maximum.

5.3 METHOD OF SECTIONS

This method is useful when we wish to know the forces in just a small part of a structure. It is of some value in checking the results of a fuller analysis. In design work it may occur when a repair is to be specified. The examples chosen here not only show the method but improve our understanding of structures.

We wish to find the greatest change of force in member X of the structure in figure 5.9 as the load rolls across. The force due to structural self-weight will

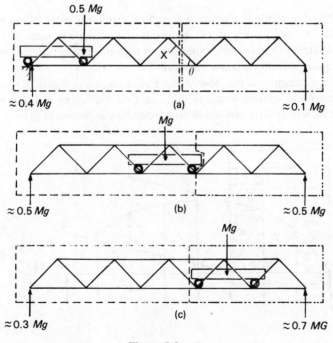

Figure 5.9

be constant so that we can ignore it in this question. We imagine the structure to be cut along the dashed line. This line is drawn round the largest convenient region (control volume) so that we can ignore various internal details. We can choose either of the two control volumes shown, the one with the least number of forces crossing the boundary; these are all we need; the internal forces within the boundary are irrelevant.

Points to note are:

(1) In figure 6.9a only half the load is on the bridge. The other half goes directly to the support;

The horizontal forces need not be considered; by resolving vertically the force in X is found from either side, whichever is easier.

Thus the forces are

In (a) consider right hand part

$$F_X \sin \theta = 0.01\, Mg \qquad F_x = 0.1\, Mg \operatorname{cosec} \theta \qquad \text{compression}$$

In (b) consider right hand part

$$F_X \sin \theta = 0.5\, Mg \qquad F_x = 0.5\, Mg \operatorname{cosec} \theta \qquad \text{compression}$$

In (c) consider left hand part

$$F_X \sin \theta = 0.3\, Mg \qquad F_x = 0.3\, Mg \operatorname{cosec} \theta \qquad \text{tension}$$

This change of mode is a notable aspect of structures with rolling loads; it is important to design for it (a) in terms of fatigue stresses, (b) ensuring adequate strength in tension especially at the end connection and also ensuring adequate stability against buckling.

Another example is the bow-string girder shown in figure 5.10; the loads could represent structure self-weight. To find the forces in J and K without solving the whole structure we draw a boundary as shown in (a).

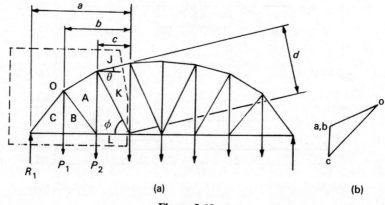

(a)   (b)

Figure 5.10

By taking moments about the KL junction we eliminate K and L for the present. The external moments on the box are zero for equilibrium

$$R_1 a - P_1 b - P_2 c - Jd = 0$$

hence J is found.

To find K we consider vertical equilibrium (this gets rid of the unknown L since it has no vertical component). We cannot readily predict whether K will be tensile or compressive. Let us assume that it is tensile. For vertical equilibrium

$$R_1 - P_1 - P_2 - J \sin \theta - K \sin \phi = 0$$

If we were wrong about K being tensile, the equation will tell us because the answer for K will in that case come out negative.

The bow-string girder is an economical form for steady, distributed loads. The triangle of forces at each junction of the top member can be as in figure 5.10b so that the diagonals are force-free, acting only as stabilisers. The limiting case is the arch, the converse is the suspension bridge or catenary; the horizontal member is replaced by the earth. Strong ground fixings are essential in these structures.

5.4 THEORETICAL OPTIMUM STRUCTURES

This and the remaining sections of chapter 5 are regarded as more advanced than the preceding.

Students often ask what angle and depth should be chosen in designing a structure. This is a compromise between available space, the cost and weight of junctions, buckling considerations, etc. If we set these considerations aside temporarily then we can find an optimum, least-weight structure for a given duty. We start with a simple bridge of negligible self-weight in comparison with the point load in the middle of a span L. The bridge rests on supports at the same level as the load and the structure must leave one half-space unobstructed (this could be as shown or upside down, depending on where we need clear space; as will be seen below, when no clearance is needed we can design a slightly lighter structure). Furthermore, we start with the simplest structural form, varying only the shape, figure 5.11. To use the material with greatest economy we must make all members as slim as possible. Each member must be so thin that it is working at the highest permissible stress, σ_{max}. This means that the cross-section of each member is made proportional to its force.

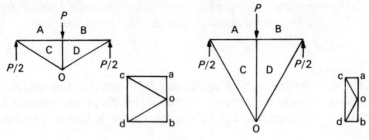

Figure 5.11

If we call the structure angle α, then

$$\text{members AC, BD are of length } \frac{L}{2} \quad \text{force} = \frac{P}{2} \times \cot \alpha$$

$$\text{members OC, OD are of length} \frac{L}{2} \sec \alpha \quad \text{force} = \frac{P}{2} \times \operatorname{cosec} \alpha$$

$$\text{member CD is of length } \frac{L}{2} \tan \alpha \quad \text{force} = P$$

From these equations and from the diagrams we see that a structure of low angle has short members but with high forces in them, while a large angle gives long members with lower forces. The extreme values are clearly un-economic, so we expect to find a best angle somewhere in between.

It is assumed that all the members are of the same material, of density ρ, and of such cross-section that they are at the same stress σ_{max}. Then the mass of each member is density × length × cross-section, or density × length × force/σ_{max}. We call $\rho/\sigma = k$

$$\text{Mass of AC} = k \, \frac{L}{2} \times \frac{P}{2} \times \cot \alpha$$

$$\text{Mass of BD} = k \, \frac{L}{2} \times \frac{P}{2} \times \cot \alpha$$

$$\text{Mass of OC} = k \, \frac{L}{2} \times \sec \alpha \times \frac{P}{2} \operatorname{cosec} \alpha$$

$$\text{Mass of OD} = k \frac{L}{2} \times \sec \alpha \times \frac{P}{2} \operatorname{cosec} \alpha$$

$$\text{Mass of CD} = k \, \frac{L}{2} \times \tan \alpha \times P$$

For ease of analysis, the angle terms are transposed into functions of h, where h is the length of member CD, that is, the maximum structure depth

$$\tan \alpha = \frac{h}{L/2} = \frac{2h}{L} \qquad \cot \alpha = \frac{1}{\tan \alpha} = \frac{L}{2h}$$

$$\sec \alpha = \frac{(h^2 + L^2/4)^{1/2}}{L/2} \qquad \operatorname{cosec} \alpha = \frac{(h^2 + L^2/4)^{1/2}}{h}$$

With these substitutions

$$\text{the total mass} = \left(\frac{PKL}{2}\right)\left(\frac{L}{2h} + \frac{2(h^2 + L^2/4)}{hL} + \frac{2h}{L}\right)$$

$$= \left(\frac{PKL}{2}\right)\left(\frac{L}{2h} + \frac{2h}{L} + \frac{L}{2h} + \frac{2h}{L}\right)$$

$$= \left(\frac{PKL}{2}\right)\left(\frac{L}{h} + \frac{4h}{L}\right)$$

To find the minimum mass, we differentiate with respect to h and set to zero

$$-\frac{L}{h^2} + \frac{4}{L} = 0$$

$$\frac{1}{h^2} = \frac{4}{L^2}$$

$$h = \frac{L}{2}$$

The second differential is positive, so confirming that we have a minimum not a maximum. Feeding this value back into the mass equation gives a total mass of $2P\rho L/\sigma$.

Proceeding to a more elaborate form, figure 5.12, we adopt a more convenient notation calling the span $4a$, the load $2R$. As before, density and

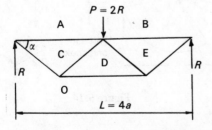
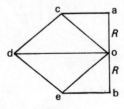

Figure 5.12

stress are common to all members hence mass is proportional to length times force. Listing the members

Members	Length	Force	Length × Force
AC+BE	$4a$	Ra/h	$4Ra^2/h$
OD	$4a$	$2Ra/h$	$4Ra^2/h$
OC+CD+DE+EO	$4(a^2+h^2)^{1/2}$	$R(a^2+h^2)^{1/2}/h$	$4R(a^2+h^2)/h$
Total			$(12a^2/h+4h)R$

Differentiating and setting to zero

$$\frac{-12a^2}{h^2} + 4 = 0$$

$$h = a\sqrt{3}$$

$$\text{Total mass} = \frac{8\sqrt{3}a\rho R}{\sigma}$$

or (using $P = 2R$, $L = 4a$)

$$\text{mass} = \frac{1.732\,PL\rho}{\sigma}$$

$$\alpha = \arctan\sqrt{3} = 60°$$

It is noticed that the optimum structures discussed so far both fit into a semicircle. Carrying this thought further leads to the structure shown in figure 5.13. Because of the increasing number of variables it is not easy to show that this is optimised by being inscribed into a semicircle; however when worked out it gives the lightest structure so far, at $1.657\,P\rho L/\sigma$. The ultimate least-weight structure for these loading conditions was shown by Michell [34] to be a semicircular rim supported by an infinite number of spokes, shown in figure 5.14.

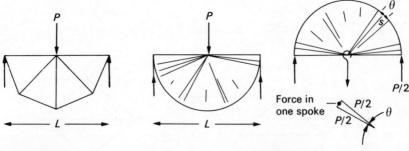

Figure 5.13   Figure 5.14

The force in the rim is $P/2$; the necessary mass of the spokes is arrived at by assuming that there are n spokes each responsible for an arc portion of angle

θ. The force in one spoke S is shown in the triangle of forces to be $P\theta/2$. Since there are n spokes, $n \times \theta$ must equal the whole semicircle angle, π

$$\text{The rim mass} = \frac{\text{density} \times \text{length} \times \text{force}}{\text{stress}} = \frac{\rho \times \pi \times L}{2} \times \frac{P}{2\sigma}$$

$$= \frac{\pi P \rho L}{4\sigma}$$

$$\text{The spoke mass} = \frac{n \times L}{2} \times \frac{\rho \times P\theta}{2\sigma}$$

since $n \times \theta = \pi$, again

$$\text{spoke mass} = \frac{\pi P \rho L}{4\sigma}$$

Thus

$$\text{the structure mass} = \frac{1.57 P \rho L}{\sigma}$$

We can obtain an even lighter structure if we disobey one of the rules and use the whole space above and below, the infinite Michell field rather than the semi-infinite. Then the lightest structure takes the form of two 90° arcs with spokes, joined to the reaction points by straight bars at 45° to the horizontal, figure 5.15. The mass of this structure is $1.285 P \rho L/\sigma$.

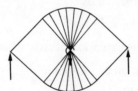

Figure 5.15

Practical optimum structures will always be shallower rather than deeper as compared with the theoretical optimum.

5.5 SIMPLE REDUNDANT STRUCTURES

The full methods for solving redundant structures are too elaborate to be given here; moreover these methods do not help to convey a sense of structure. The two methods given here are readily followed on the basis of the simple methods in this chapter. In addition, the examples convey useful structural lessons.

First we consider the bracket of figure 5.16, made of three hinged bars of equal cross-section A. A horizontal force P causes R to lengthen, T to shorten while S stays the same length, just swinging a little to the right. The forces F_R, F_T are equal in magnitude.

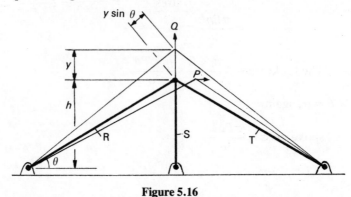

Figure 5.16

Resolving horizontally

$$P = (F_R + F_T) \cos \theta$$

hence $F_R = F_T = \frac{1}{2}P \sec \theta$

Next we put on a vertical force Q, extending the bracket upwards by an amount y. Resolving vertically

$$Q = F_R \sin \theta + F_S + F_T \sin \theta \qquad (5.1)$$

The forces are found from the extensions: force = stress $\times A$ = strain $\times E \times A$

$$F_R = F_T = \frac{EAy \sin \theta}{(h/\sin \theta)} = \frac{EAy \sin^2 \theta}{h}$$

$$F_S = \frac{EAy}{h}$$

Substituting in 5.1 gives

$$Q = \frac{EAy (1 + 2 \sin^3 \theta)}{h}$$

$$F_S = \frac{Q}{(1 + 2 \sin^3 \theta)}$$

$$F_R = \frac{Q \sin^2 \theta}{(1 + 2 \sin^3 \theta)}$$

The structural lesson is that the most direct, stiffest load path receives the greatest share of the load.

A more common and more useful redundant structural form is the cross-braced panel. The diagonal braces are usually made of equal cross-section and we are about to show that they tend to share the load fairly evenly; when designed accordingly they use little extra material compared with an equivalent single brace and form a useful safety feature. As shown in figure 5.17a, should a misfortune befall any one member the structure is still fully triangulated and capable of standing about half the original peak load which will often be adequate until a repair can be made (provided the design is safe from buckling).

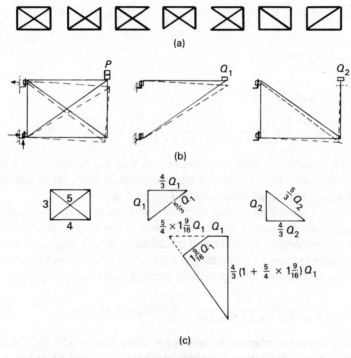

Figure 5.17

The method recommended for finding the forces in such a structure consists of splitting it into two sub-structures and working out the loads needed to cause a given deflection. The way it works out varies slightly with the detailed load and reaction conditions. As a worst case we consider a gate loaded at the top of one post and hinged so that all the *vertical* force is taken by the bottom hinge. This gives the greatest inequality between the substructures. Figure 5.17b shows the method, then in (c) we show a particular example.

Let us assume a gate 1.5 m high, 2 m wide, made of wood, $E = 13.33 \times 10^9$

N/m^2, all members of 20 cm² cross-section. We shall estimate the deflection under a load P of 1 kN and the forces in the diagonals.

The first sub-structure under load Q_1 will have a tension T_1 and a compressive force C_1. The triangle of forces has the same shape as the structure, a 3,4,5 triangle.

$$T_1 = Q_1 \times \frac{4}{3}$$

The member will extend by

$$Q_1 \times \frac{4}{3} \times \frac{2}{20 \times 10^{-4} \times 13.33 \times 10^9} = Q_1 \times 10^{-7} \, m$$

$$C_1 = Q_1 \frac{5}{3}$$

The member will shorten by

$$Q_1 \times \frac{5}{3} \times \frac{2.5}{20 \times 10^{-4} \times 13.33 \times 10^9} = 1.5625 Q_1 \times 10^{-7} \, m$$

No other members are involved, hence the downward deflection = $3.9375 Q_1 \times 10^{-7}$ m obtained from the vector triangle in figure 5.17c (as in section 5.1). There is also a horizontal deflection of $Q \times 10^{-7}$ m.

The second sub-structure has three components of downward deflection: the triangle, shortening of the vertical between the hinges which lowers the position of the sloping member and incidentally gives a slight extra slope to the lower member, and a shortening of the upright directly under the load Q_2.

These latter two shortenings under a force Q_2 are each

$$\frac{Q_2 \times 1.5}{20 \times 10^{-4} \times 13.3 \times 10^9} = 0.5625 Q_2 \times 10^{-7} \, m$$

The deflection of the triangle, by analogy with the first, = $3.9375 Q_2 \times 10^{-7}$ m.

The horizontal deflection is unimportant since the right hand member is free to pivot, thus we are entitled to say that the vertical deflections of both substructures must be common. Hence

$$3.9375 Q_1 = (3.9375 + 2 \times 0.5625) Q_2 \qquad \text{or } Q_1 = 1.286 Q_2$$
$$\text{The total load } P = Q_1 + Q_2 = 2.286 Q_2$$

Thus if $P = 1000 \, N$, $Q_2 = 1000/2.286$ N and the deflection under P of 1 kN is

$$\frac{(3.9375 + 2 \times 0.5625) \, 1000}{2.286 \times 10^{-7}} \text{ metres} = 0.22 \, mm$$

In a real wooden gate there would be further deflections in the joints, at the hinges, etc.

In practice, redundant structures are often cross-braced symmetrically (as in figure 5.17). These can be treated simply by pretending that one cross-brace has the cross-section of both and the other one is absent. This is also valid for pre-tensioned cross-braces by working in terms of force *changes* from the original condition, up to the point when one of the braces buckles. Then the buckled member is treated as a force input of magnitude P_{ce} regardless of its deflection; the other cross-brace is then single and non-redundant.

For the rare case of an unsymmetrical structure with one redundant member, the theorem of lowest strain energy can be used. This assumes that the structure will take up that configuration which contains the lowest strain energy. The method is to remove one member, replacing it with a tensile force Q in the direction of the member. Then we find the forces in all the remaining members due to the loads applied to the structure including the unknown force Q, separately, using Bow's notation.

Next we have to assume provisional values for the cross-sections of all the members. These can be actual values or they can be those values which will in the end give equal stresses (this makes it easier). From the total force in each member we find the strain energy. The total sum of strain energy, including member Q, is differentiated with respect to Q and set to zero. This gives an answer for the force Q.

A variation of this method is to remove one member as before, put in a unit force in the direction of the removed member and find the stiffness of the structure with respect to that direction of loading. Then we choose a provisional member cross-section and find its stiffness. Equating the deflections gives the actual force in the removed member. This method is suitable for cases where the redundant member is of incorrect length; the discrepancy is allowed for when comparing the deflections.

The kind of redundant structure found in practice is more like the body of a coach. A motor coach or a railway coach consists of three beams in parallel, like a leaf spring stack of equal-length leaves: the chassis, the body sides and the roof. The window pillars have a more dubious role, a mixture of tension member and beam. A modern solution would use finite element methods. By hand, as a credibility check, one would assume a deflected form of arbitrary magnitude but realistic shape. Then the strain energy in all components would be added up and set equal to the potential energy sum of the loads and deflections; the loads are known, the deflection at each bit of load is taken as the assumed deflection at this point times a factor k. Hence the final deflections everywhere are simply k times the assumed deflection, where k is the only unknown in the equation. From the deflection thus found, the stresses are obtained.

PROBLEMS

5.1 Estimate the ratio of the tensions T_1, T_2 for the two rigs shown in figure 5.18a (upright and raked mast). You may use a mixture of drawn and calculated values.

Ans. $T_2/T_1 = 3.75, 3.68$.

5.2 Figure 5.18b shows a lifting lug bolted to a component. Assuming friction to be small find the force P which will cause an equivalent tensile stress in the 15 mm shank, at angles of $\theta = 0°, 45°, 90°$. Refer to section 2.6 and assume steady loading. Assume uniform shear, for simplicity's sake.

Ans. 35.3, 17.7, 15.8 kN.

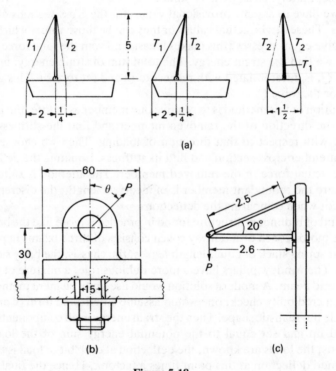

Figure 5.18

5.3 Figure 5.18c shows a pillar crane. The horizontal bar is a steel tube 50 mm outer diameter (o/d), 40 mm inner diameter (i/d). The sloping steel tie of 30 mm diameter is threaded with an M30 thread, root diameter 25.2 mm over a short distance only $E = 200\,000$ N/mm². Find the download which may be carried at P if the stress is not to exceed 120 N/mm² ignoring stress concentration. Estimate the deflection at this loading.

Ans. 20.5 kN, 6 mm (approx.) down, 1 mm →.

5.4 A bracket as in figure 5.16 is made of steel bars of 200 mm² cross-section, $\theta = 35°$. Find the displacement if a load in the Q direction is applied such that the highest stress is 100 N/mm² ignoring stress concentrations at joints, etc. Also find the stress in the side-bars (S, T). $h = 160$ mm, $E = 200\ 000$ N/mm².

Ans. 0.08 mm, 27.5 kN, 35 N/mm².

5.5 A foot-bridge is represented by figure 5.19a. If all dimensions are in metres, the walkway is considered as hinged at every junction and carries a load of 100 kg/m, find the forces in A and B.

Ans. 3468, 3533 N.

5.6 A telephone pole of 200 kg mass is being raised by using poles as in figure 5.19b. Find the force T required in the rope to start lifting, also the force when the middle of the load is about 2 m above ground.

Ans. 6.1 kN, 5 kN.

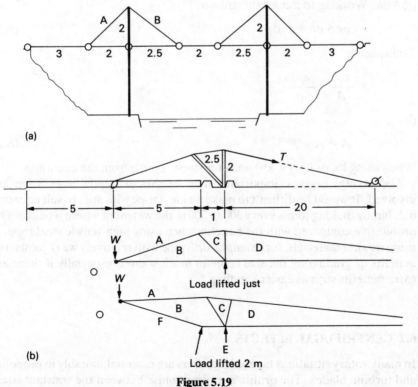

(a)

(b)

Load lifted just

Load lifted 2 m

Figure 5.19

6 MORE ADVANCED TENSION PROBLEMS
(Second to third year standard)

6.1 GRAVITY EFFECT

The weight of a member can cause appreciable stress. For example, a mine hoisting rope of 50 mm diameter, 500 m long, weighs some 5 to 6 tons. The stress due to this weight is something like 40 N/mm² on top of the duty stresses due to load, inertia and curvature. When we consider tethering a weather balloon, dredging samples from the sea bed or drilling deep holes, we need to consider tailoring the rope or drill pipe to the duty in more detail.

If the material density $= \rho$, the cross-section A, then ideally we should adjust the cross-section to suit the load. A piece of height dh exerts a force $\rho g A \, dh$. Working to a constant stress σ

$$\rho g A \, dh = \sigma \, dA \tag{6.1}$$

Transposing

$$\frac{dA}{A} = \frac{\rho g \, dh}{\sigma}$$

or

$$A = A_0 e^{\rho g h / \sigma} \tag{6.2}$$

When using lb_f or kg_f ($= kp$) units for stress, omit g from the equations.

The savings become important when heights or depths over 5 km are involved. It would be difficult to make an ideal rope with area to suit equation 6.2, but by making joints every 300 to 500 m the weight of a joint would not be prohibitive compared with the benefits when using high-tensile steel rope. If using weaker materials, for example, drill pipe with relatively weak joints, the benefits of graduating the size come in much sooner, especially if there are extra benefits such as easier handling.

6.2 CENTRIFUGAL EFFECTS

In many rotary situations tapered members are essential, notably in propeller and turbine blades. The profile is a compromise between the constant stress form (see below) and the needs of stiffness and aerofoil shape. In slower machines, constant area is sufficient; well-known examples are the chain flail manure spreader, developed from the mine exploder; also one form of grass cutter.

The general expression for a rod rotating about a point O at ω radians per second, of density ρ and cross-sectional area A, carrying an end mass M as in figure 6.1, that is, force at a general radius r, between r_0 and R

$$F_r = M\omega^2 R_M + \int_r^R \rho A \omega^2 r \, dr \qquad (6.3)$$

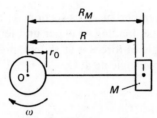

Figure 6.1

If using gravitational units such as lb_f or kp (kg_f), the g factor must be included in equations 6.3 to 6.5 by using ρ/g. In inch units, $g = 32.174 \times 12 = 386$ in/sec^2, not 386.4 as often reported, including reference 5.)

For design purposes the end mass M and the permissible stress σ decide the minimum area A_R at the outer radius; at $r < R$ we must make the area larger than minimum to resist the additional forces due to the rod itself. The ideal area increase is that which only just compensates for the extra force, thus maintaining a constant stress; this follows from the equation

$$A = A_0 e^{-\frac{\rho\omega^2 r^2}{2\sigma}} \qquad (6.4)$$

Derivation of equation 6.4: differentiating equation 6.3 gives $dF/dr = \rho A \omega^2 r$

$$A = \frac{F}{\sigma}$$

hence

$$\frac{dA}{dr} = \frac{dF}{dr} \,/\sigma \text{ for constant } \sigma$$

hence

$$\frac{dA}{A} = \frac{\rho\omega^2 r \, dr}{\sigma}$$

hence

$$\ln A = \frac{\rho\omega^2 r^2}{2\sigma} + \text{constant}$$

For a uniform member with no end load, the greatest stress is at the fixing (r_0) and is

$$\sigma_{max} = \tfrac{1}{2}\rho\omega^2 \left(R^2 - r_0^2\right) \tag{6.5}$$

6.3 CURVED MEMBERS WITH TRANSVERSE FORCES (FOR EXAMPLE, BELTS, WHEEL RIMS, PIPES, ETC.)

Figure 6.2 shows a member of cross-section A, curved to a radius r and subjected to a transverse loading of w per unit length. An element dy is in equilibrium if its triangle of forces is as shown, producing a tensile stress σ. The element subtends an angle dθ at O.

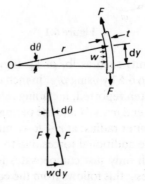

Figure 6.2

From the triangle of forces

$$w\,dy = F\,d\theta = A\sigma\,d\theta$$

from geometry

$$dy = r\,d\theta$$

hence

$$\text{force} = A\sigma = wr$$

$$\text{stress } \sigma = \frac{wr}{A} \tag{6.6}$$

In a flywheel rim or in a moving belt the loading w is due to centripetal acceleration v^2/r; the mass of an element $= \rho A\,dy$ while the tensile force in the rim $= A\sigma$ thus from the triangle of forces

$$A \sigma \, d\theta = \frac{\rho A v^2 \, dy}{r}$$

$$\sigma = \rho v^2 \tag{6.7}$$

Vessels under fluid pressure are very similar. Considering unit width normal to the paper, the load w per unit length becomes p, the conventional symbol for force per unit area (fluid pressure).

From balance of forces, unit width

$$F \, d\theta = p \, dy = pr \, d\theta$$

$$\text{Stress } \sigma = F/t$$

$$\sigma = \frac{pr}{t} \tag{6.8a}$$

This is called tangential, or commonly hoop stress, as distinct from axial or radial stress.

For a spherical shell we can no longer consider unit width since a similar force F exists normal to the paper. We consider a square patch $dy \times dy$.

The forces are $p \, dy^2$ to the right, opposed by $2F \, d\theta$ to the left, one F in the paper plane and another at right angles, since figure 6.2 applies equally to elevation or plan view. The stress $\sigma = \text{force/area} = F/(t \, dy) = F/(tr \, d\theta)$. From force balance

$$p \, dy^2 = p(r \, d\theta)^2 = 2F \, d\theta = 2\sigma tr \, d\theta^2$$

$$\sigma = \frac{pr}{2t} \tag{6.8b}$$

Pipes and pressure vessels are treated in chapter 9. Here we consider more general cases.

A common case is a ring shrunk on to a solid shaft or on to another ring by thermal contraction, or pressed together. The two members exert a mutual radial pressure p, giving rise to hoop stresses σ_h. The simplest way to discuss this is by examples.

Example 6.1

A steel ring gear of 300.0 mm bore is fitted on to a rigid flywheel of 300.2 mm diameter.

(a) What temperature must the ring be heated to in order to expand it to 300.5 mm for ease of fitting? Let $\alpha = 11 \times 10^{-6}/°C$

At this point it may be wise to dispose of a popular fallacy. It is sometimes believed that when a ring expands due to temperature, the hole gets smaller. This seems to derive from experience with swelling of wood: if a block of wood with a hole is wetted, the hole tends to become smaller. The reason is that such swelling starts at the surface; the swelling action slows down the penetration of wet to the interior. If metal is heated slowly, thermal conduction ensures that the growth of thickness is not much faster than that of the periphery. Once the temperature is uniform, all dimensions expand in the same ratio.

$$\frac{\text{expansion}}{\text{original length}} = \text{temperature change} \times \frac{\text{coefficient of linear}}{\text{expansion}}$$

$$\frac{0.5}{300} = \theta \times 11 \times 10^{-6}$$

$$\theta = \frac{500\ 000}{3300} = 150\ ^\circ\text{C}$$

(b) What is the stress in the ring when installed, ignoring stress concentration?

From the tensile laws

$$\text{stress} = \text{strain} \times \text{Young's modulus}$$

$$\sigma_h = \frac{0.2\ \pi}{300\ \pi} \times 200\ 000\ \text{N/mm}^2 = 133\ \text{N/mm}^2$$

When neither member can be called rigid relative to the other, both elasticities need considering.

Example 6.2

A steel railway wheel of 800 mm diameter has a rim of effective cross-section 3600 mm². Onto this is shrunk an outer rim of 798.8 mm bore, 2400 mm² effective cross-section, 80 mm wide at the interface. Find the stress in each rim and the contact pressure *p*.

Using equation 6.6, *w* is the radial force per unit length = 80 *p*

$$\sigma_1 = \frac{w \times 400}{3600} \qquad \sigma_2 = \frac{w \times 400}{2400} \qquad \text{(sign convention ignored, direction obvious)}$$

Elastic contraction of inner diameter $= \dfrac{800\sigma_1}{E}$

Elastic expansion of outer diameter $= \dfrac{800\sigma_2}{E}$

Discrepancy (interference) $= \dfrac{(\sigma_1 + \sigma_2) \times 800}{E} = 1.2\ \text{mm}$

$$\sigma_1 + \sigma_2 = \frac{1.2 \times 200\ 000}{800} = 300\ \text{N/mm}^2$$

From first equation, $\sigma_2/\sigma_1 = 1.5$
therefore

$$\sigma_1 = 120 \text{ N/mm}^2 \qquad \sigma_2 = 180 \text{ N/mm}^2$$

$$w = 9\sigma_1 \qquad p = \frac{9 \times 120}{80} = 13.5 \text{ N/mm}^2$$

An important topic is differential thermal expansion with interference. The following example is particularly instructive.

Example 6.3

A bronze bush of 40 mm outside diameter, 3 mm wall thickness is lightly pressed into a steel housing 6 mm thick. The bush is a working fit against a steel gudgeon pin. For the bronze

$$\alpha = 18 \times 10^{-6}/°C \qquad E = 125\,000 \text{ N/mm}^2$$

For the steel

$$\alpha = 11 \times 10^{-6}/°C \qquad E = 200\,000 \text{ N/mm}^2$$

Find the stresses in bush and housing when the temperature is raised from 20 °C as fitted to a working temperature of 170 °C; also find the change of working clearance between bush and gudgeon pin.

Relative expansion = $40 \times 150 \times (18 - 11) \times 10^{-6} = 0.042$ mm. Per mm width, area = thickness, therefore

$$\sigma_1 = \frac{wr}{3} \qquad \sigma_2 = \frac{wr}{6} \qquad \text{from equation 6.6}$$

therefore

$$\sigma_2 = \tfrac{1}{2}\,\sigma_1$$

$$\text{Contraction of inner} = \frac{40\sigma_1}{E_1}$$

$$\text{Expansion of outer} = \frac{40\sigma_2}{E_2} = \frac{40\,(\tfrac{1}{2}\sigma_1)}{E_2}$$

$$0.042 = 40\left(\frac{\sigma_1}{125\,000} + \frac{\tfrac{1}{2}\,\sigma_1}{200\,000}\right)$$

$$\sigma_1 = \frac{0.042 \times 10^6}{40 \times 10.5} = 100 \text{ N/mm}^2$$

$$\sigma_2 = \tfrac{1}{2}\,\sigma_1 = 50 \text{ N/mm}^2$$

Clearance calculation: Neglecting radial thickness change of the bush for the moment, the bush bore will increase because of temperature but decrease because of compressive stress. The gudgeon pin is not under radial stress of any serious magnitude, so it only expands because of temperature

$$\text{Strain of bush} = \alpha\theta - \frac{\sigma}{E} = 18 \times 10^{-6} \times 150 - \frac{100}{125\,000}$$

$$= 1.9 \times 10^{-3}$$

$$\text{Strain of pin} = \alpha\theta = 150 \times 11 \times 10^{-6} = 1.6 \times 10^{-3}$$

$$\text{Clearance change} = \text{diameter} \times \text{net strain} = 34 \times (1.9 - 1.6)$$
$$\times 10^{-3} = 0.01 \text{ mm}$$

$$\text{Thickness expansion} = 3 \times 150 \times 18 \times 10^{-6} = 0.0081 \text{ mm}$$
$$\text{Net change} = 0.002 \text{ mm}$$

Comments:

(1) the use of the word strain for thermal expansion per unit length is usual in some books. It is used for brevity's sake although misleading in the colloquial sense.

(2) note how the stress due to the press fit helps to reduce the change of clearance. If we allow for Poisson's ratio, clearance change ≈ 0.

6.4 STRESSES AND STRAINS IN DISCS

In discs as opposed to thin rings, the hoop stress is far from uniform, tending to be highest at the central hole or solid centre. The important cases are

(1) a uniform rotating disc, for example, a circular saw (see section 6.4.1);

(2) a contoured disc thicker in the centre, giving constant stress (see section 6.4.2);

(3) a disc or thick pipe under a radial pressure (see section 6.4.3).

The full equations are rather forbidding and are quoted in appendix A.6. The main forms needed for design are relatively manageable. Figure 6.3 shows the notation for a whole disc and a segment to help in visualising how the equation comes about.

Taking all stresses as positive when tensile, the angular velocity as ω and density ρ, the forces on the segment are

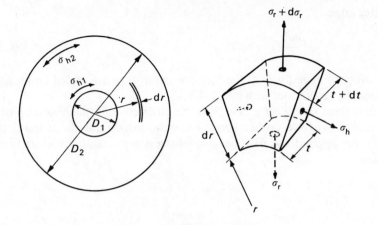

Figure 6.3

Outwards

$$(\sigma_r + d\sigma_r)(t + dt)(r + dr)\theta + (\rho\omega^2 r)r\theta t\, dr$$

Inwards

$$\sigma_r tr\theta + \sigma_h t\theta\, dr$$

Cancelling θ we can use the shortened form

$$(\rho\omega^2 r^2 - \sigma_h)t\, dr = d\,(tr\sigma_r) \tag{6.9}$$

The solutions involve reconciling the strains with use of Poisson's ratio.

6.4.1 Constant Thickness Rotating Disc

For a uniform disc of constant thickness and zero edge loading, the hoop stresses are

$$\sigma_{h1} = \frac{\rho\omega^2}{16}\left[(3 + \nu)D_2{}^2 + (1 - \nu)D_1{}^2\right] \tag{6.10a}$$

and

$$\sigma_{h2} = \frac{\rho\omega^2}{16}\left[(1 - \nu)D_2{}^2 + (3 + \nu)D_1{}^2\right] \tag{6.10b}$$

If the disc is solid (no hole), then the solution is not valid; see appendix A.6 for reasons.

$$\sigma \text{ at the centre} = \sigma_h = \sigma_r = \frac{\rho\omega^2}{32}\,(3 + \nu)D^2 \tag{6.11a}$$

at outer edge

$$\sigma_h = \frac{\rho\omega^2}{16}(1 - \nu)D^2 \tag{6.11b}$$

The effect of edge loads at the inner or outer edge are added on by super-position, in accordance with case 6.4.3.

The highest stresses are always at the inner edge if thickness is constant. The reason why we bother about the outer edge (and about the general radius, see appendix A.6) is because of stress concentrations which may be present.

6.4.2 The Constant Stress Disc

This solution is not given in many textbooks which is a pity and a sign of an impractical attitude. It dates back at least to De Laval in the nineteenth century and is the basis of practically all high speed turbine disc designs.

Its profile is derived very readily from the segment equation 6.9. If the stresses are constant in both directions throughout, which implies that the disc must be solid in the centre, then we put $\sigma_h = \sigma_r = \sigma$ into equation 6.9

$$d\,(rt\sigma) - \sigma t\,dr = \rho\omega^2 r^2 t\,dr$$

$$r\,dt + t\,dr - t\,dr = \frac{\rho\omega^2 r^2 t\,dr}{\sigma}$$

$$r\,dt = \frac{\rho\omega^2 r^2 t\,dr}{\sigma}$$

$$\frac{dt}{t} = \frac{\rho\omega^2 r\,dr}{\sigma}$$

Integrating

$$t = t_0\, e^{-\frac{\rho\omega^2 r^2}{2\sigma}} \tag{6.12}$$

This expression does not give the actual thickness; this is decided by the outer edge conditions. In most cases there is an edge load due to a ring of blades; the disc is made thick enough at the junction with the blade ring to stand the stress there, then using the equation ensures the same stress everywhere. If there is a temperature gradient then the shape can be modified. Also if a centre hole is needed, this raises the stress by 2:1 or more and calls for reinforcing material in the form of a wide hub, see figure 6.4. If there is no definite edge load, the stress tails off towards the edge.

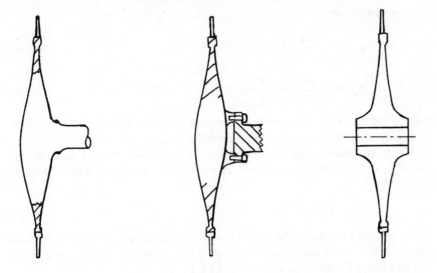

Figure 6.4

6.4.3 Uniform Disc or Cylinder

Strictly this does not apply to cylinders because their expansion is restrained by the Poisson effect lengthwise; in a disc this is absent. In a cylinder the axial expansion or shortening must be uniform at all radii, whereas the hoop and radial stresses vary. For practical purposes we often forget the difference, treating the cylinder as if it were a disc of constant thickness.

Since in most cases the radial stress is a fluid pressure, for convenience we give the stresses in terms of a fluid pressure p. When there really is a radial tension simply remember that $\sigma_r = -p$.

Internal pressure

$$\sigma_{h1} = \frac{p(D_2^2 + D_1^2)}{D_2^2 - D_1^2} \tag{6.13a}$$

$$\sigma_{h2} = \frac{2pD_1^2}{D_2^2 - D_1^2} \tag{6.13b}$$

Most books fail to notice that the pressure as such combines with the highest tensile stress σ_{h1}, giving a combined stress

$$p + \sigma_1 = \frac{2pD_2^2}{D_2^2 - D_1^2} \tag{6.13c}$$

External pressure

$$\sigma_{h1} = \frac{-2pD_2^2}{D_2^2 - D_1^2} \tag{6.14a}$$

$$\sigma_{h2} = \frac{-p(D_2^2 + D_1^2)}{D_2^2 - D_1^2} \tag{6.14b}$$

Interaction between the *external* pressure and compressive hoop stress is unimportant, not giving rise to any critical conditions.

Within the elastic region these stresses may be superposed, likewise the deflections. The radial expansions, expressed as diameter change, follow the general law

$$\frac{\Delta}{D} = \frac{(\sigma_h - \nu\sigma_r)}{E} \tag{6.15}$$

In the constant thickness disc under centrifugal stress alone, radial stresses are absent at the inner and outer edges, so $\Delta_1/D_1 = \sigma_{h1}/E, \Delta_2/D_2 = \sigma_{h2}/E$ the stresses being found by equation 6.10 or 6.11.

The constant stress disc is simpler, $\Delta/D = (1 - \nu)\sigma$.

Thick cylinders or their equivalents follow equation 6.15 of course; the stresses are found by equation 6.13, 6.14 or both if necessary, then carefully noting the signs we apply equation 6.15. The sign convention is assisted by common sense. Tensile stress enlarges the object, a radial pressure pushes the relevant surface away.

In the case of superposed systems, either use equation 6.15 on the superposed stress basis or refer to the equations given in appendix A.6 which do the job for you.

Example 6.4

An electric motor has laminations of density 7800 kg/m³, $\nu = 0.28$, as shown in figure 6.5. The 24 slots have a semicircular base and carry windings of 1.1 kg

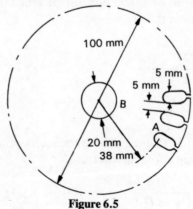

Figure 6.5

total mass at an effective radius of 42 mm. The stack of laminations is 100 mm high normal to the paper. Find the tensile stress at A and B at 12 000 rpm. With inertia forces stick to N m units.

(1) Estimate mass of laminations between slots:

$$\text{gross area} = \pi(0.05^2 - 0.0355^2) = 0.0039 \text{ m}^2$$

Slot area just under half

$$\text{Mass} = 7800 \times 0.0019 \times 0.1 = 1.5 \text{ kg}$$

(2) Find centripetal force

$$(1.1 + 1.5) \times \left(\frac{2\pi \times 12\ 000}{60}\right)^2 \times 0.042 = 172\ 400 \text{ N}$$

(3) Tensile area at radius $A = 24 \times 5 \times 100 \text{ mm}^2 = 12\ 000 \text{ mm}^2$;
(4) Stress concentration factor from figure 4.17 = 1.59;
(5) Stress at A = 22.9 MN/m²;
(6) Find σ_{r2} by taking centripetal force evenly over base of slot circle

$$\sigma_{r2} = \frac{172\ 400}{\pi \times 0.071 \times 0.1} = 7.73 \text{ MN/m}^2$$

(7) From equations 6.10a and 6.14a

$$\sigma_{hl} = 12.9 + 16.9 = 29.8 \text{ MN/m}^2$$

Example 6.5 (Advanced standard)

A fan rotor as in figure 6.6 runs at 1432 rpm (150 radians/s). It is made of titanium alloy, density 4500 kg/m³, $E = 1.1 \times 10^{11}$ N/m², $\nu = 0.33$. The shroud S and blades B amount to a 7 mm increase of plate thickness between XX and the outer edge but without adding any support. Neglecting any bending effects find the highest tensile and shear stress in the rotor and the increase of diameter at full speed. This is taken from an actual case study with slight simplification (see section 10.4).

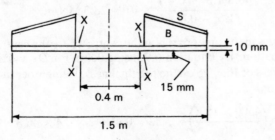

Figure 6.6

Consider the inner and outer parts separately, exerting a radial tensile stress σ_x on each other at the junction, the stress acting on the 10 mm thickness so that the thicker portion has a radial stress of $\sigma_x \times 10/25 = 0.4\sigma_x$ except just around the weld. The radial expansion Δ of the inner and outer parts must be equal at x. Inner part:

From equations 6.15, 6.11b, 6.14b (or A.5)

$$\frac{\Delta}{0.4} = \left(\frac{4500 \times 150^2}{16E}\right) \times 0.67 \times 0.4^2 + \frac{0.4 \times \sigma_x \times 0.67}{E}$$

$$= \frac{0.678 \times 10^6 + 0.268\,\sigma_x}{E}$$

Outer part:

To allow for the load due to shroud and blades, multiply the density by $(10 + 7)/10 = 1.7$. Then use equations 6.10a, 6.13a or A.4 (Δ is at the inner edge)

$$\frac{\Delta}{0.4} = \left(\frac{1.7 \times 4500 \times 150^2}{16E}\right)(3.33 \times 1.5^2 + 0.67 \times 0.4^2)$$

$$- \frac{\sigma_x[(1.5^2 + 0.4^2)/(1.5^2 - 0.4^2) + 0.33]}{E}$$

$$= \frac{81.76 \times 10^6 - 1.483\sigma_x}{E}$$

The expansions Δ must be equal
Therefore

$$81.76 \times 10^6 - 1.483\sigma_x = 0.678 \times 10^6 + 0.268\sigma_x$$

$$\sigma_x = 46.3 \text{ N/mm}^2$$

The greatest stress in the outer part may be the radial stress σ_x or the hoop stress by equations 6.10a and 6.13a

$$\sigma_{hl} = \left(\frac{1.7 \times 4500 \times 150^2}{16}\right)(3.33 \times 1.5^2 + 0.67 \times 0.4^2)$$

$$- \frac{46.3(1.5^2 + 0.4^2)}{(1.5^2 - 0.4^2)} \times 10^6 = 28.3 \text{ MN/m}^2$$

The greatest stress in the inner part is at the centre where the centrifugal effect adds its contribution. By equation 6.9, using 0.4 m for D_2 and zero for D_1, with an edge stress of $10\sigma_x/25$ on the full thickness. Remember that for a solid disc $\sigma_{hl} = \sigma_{rl} = \sigma_0$

$$\sigma_0 = \left(\frac{4500 \times 150^2}{16}\right) \times 3.33 \times 0.4^2 + 2 \times 0.4 \times 46.3 \times 10^6 - \sigma_0$$

$$\sigma_0 = 20.2 \text{ N/mm}^2$$

The highest shear stress is $\frac{1}{2}\sigma_x$ acting at 45° to the disc plane. By equation 6.15, $\Delta_2 = 0.2$ mm at the rim (remember the extra loading!)

Example 6.6 (Advanced standard)

A turbine disc as in figure 6.7 has 60 blades of 22 g each and 60 fastening teeth of 15 g each, acting at an effective diameter of 300 mm. The other dimensions are as in the figure. If the disc is to run at 25 000 rpm and is made of material of 8 g/ml density, $\nu = 0.28$, find the thicknesses t_0 and t_1 such that the stress $\leq$ 300 MN/m². Ignore stress concentrations.

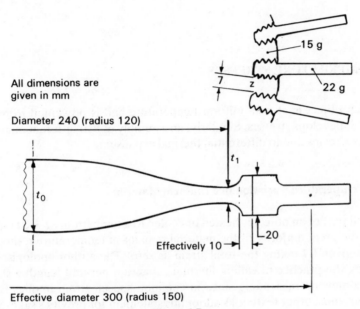

All dimensions are given in mm

Diameter 240 (radius 120)

Effective diameter 300 (radius 150)

Figure 6.7

First check the stress at z

$$\text{Force} = m\,\omega^2 r = (0.022 + 0.015)\left(\frac{25\ 000 \times 2\pi}{60}\right)^2 \times 0.15$$

$$= 38\ 039\ \text{N}$$

Area $= 7 \times 20$ mm², stress $= 272$ N/mm².
Balance of forces over angle θ:rim stress $= E \times$ rim strain Δ/D, see equation 6.18

$$\text{Outward force} = 38\ 039 \times 60 \times \frac{\theta}{2\pi}$$

$$\text{Inward forces} = 20 \times 10 \times (1 - 0.28) \times 300 \times \theta \text{ for rim}$$

$$+ t_1 \times 120 \times \theta \times 300 \text{ N}$$

Cancelling θ

$$t_1 = \frac{363\ 245 - 43\ 200}{36\ 000} = 8.9 \text{ mm}$$

From equation 6.12

$$t_0 = 8.9 \text{ e}^{\frac{\rho\omega^2 r^2}{2\sigma}} = 8.9 \text{ e}^{1.316} = 33.2 \text{ mm}$$

6.5 TEMPERATURE STRESSES

When an object is at non-uniform temperature and is prevented from distorting, it develops stresses. Similarly, an assembly of two different materials contains stresses due to differential thermal expansion.

6.5.1 Temperature Variation in a Uniform Material

If a small part of an object is heated or cooled to a temperature $t + \Delta t$ rapidly so that the great majority of the material remains at temperature t, then the stress is given by taking the total strain as zero. The author apologises for following the practice of calling thermal expansion per unit length a strain; this is somewhat misleading since in common speech strain implies force. However, most other textbooks adopt this statement for brevity's sake. Thus total strain

$$\epsilon = \frac{\sigma}{E} + \alpha\Delta t \tag{6.16}$$

In the case of a very local difference, $\epsilon = 0$, $\sigma = -E\alpha\Delta t$.
    Steady state heat-flow follows the law

$$\text{heat flow} = \text{area} \times \text{temperature gradient} \times \text{conductivity}$$

In a pipe with radial heat flow, a very common engineering case

$$Q = -2\pi rLk\frac{dt}{dr} = \text{constant}$$

the negative sign showing that heat flows *down* the temperature gradient. This integrates to give

$$t - t_1 = -\left(\frac{Q}{2\pi Lk}\right)\ln\left(\frac{r}{r_1}\right)$$

(6.17)

In a pipe the radial expansion is quite complicated if the wall is thick compared with the inner diameter as we have seen in section 6.4, giving highest stress at the inner surface. Without going into extensive details, the stress distribution is as in figure 6.8a. Most pipes are thin enough for the departure from a straight line to be no greater than the general uncertainties of temperature prediction. In those cases the stress is taken as proportional to half the temperature difference from inside to outside

$$\sigma = \frac{\alpha E(t_1 - t_2)}{2}$$

(6.18)

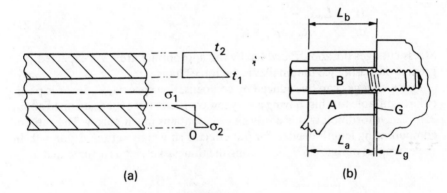

(a)   (b)

Figure 6.8

6.5.2 Uniform Temperature, Various Materials

This is very common, particularly in the form of an aluminium, magnesium or other high-expansion alloy secured with steel bolts. Figure 6.8b shows such a case; a member A is secured by bolts B, possibly with a thin gasket G. As assembled, at uniform temperature, the bolt is tightened to a force P which automatically causes a compressive force $-P$ in the other parts. This is taken as the starting point for the displacement calculations.

The effect of a temperature change dt is a displacement dx at the bolt-head/cover contact face. Since dx is common to both parts but the expansion coefficients differ, the force P has changed by an amount dP. Our object is to find this amount so that we can find the stress changes.

From the bolt's viewpoint

$$dx = L_b\alpha_b\,dt + \frac{dP\,L_b}{A_b E_b}$$

The compressed parts expand due to the temperature change but shorten due to the force dP

$$dx = L_a\alpha_a\,dt - \frac{dP\,L_a}{A_a E_a} + L_g\alpha_g\,dt - \frac{dP\,L_g}{A_g E_g} \quad \text{(note the sign)} \quad (6.19)$$

For neatness of writing, L/AE is lumped together as the compliance or resilience of the component, called R. It is the reciprocal of the stiffness k. The resilience will be found useful again in the next section. In bolts, the threaded portion must be taken into account, being appreciably thinner than the main shank.

Equating dx gives

$$(L_b\alpha_b - L_a\alpha_a - L_g\alpha_g)\,dt = -(R_b + R_a + R_g)\,dP \qquad (6.20)$$

$$\frac{dP}{dt} = \frac{L_b\alpha_b - L_a\alpha_a - L_g\alpha_g}{R_b + R_a + R_g} \qquad (6.21)$$

Stress changes $d\sigma$ and dP divided by the appropriate area. As expected, the greater the resilience the smaller is the force change.

Force changes can be reduced by compensating spacers, C. Invar is particularly suitable for this, having a very low coefficient of expansion up to about 250 °C. Unfortunately it is costly since it contains much nickel. The ratio of compensating length needed for full effect is shown by setting $dP/dt = 0$. In the examples of figure 6.9, neglecting the thin gasket, let us assume that $\alpha_b =$

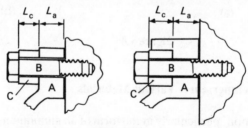

Figure 6.9

11.5×10^{-6} (carbon steel), $\alpha_a = 23 \times 10^{-6}$. (For stainless steel bolts α_b is greater, so less compensation is needed.)

From 6.21 modified, $\alpha_b L_b - \alpha_a L_a - \alpha_g L_g - \alpha_c L_c = 0$ (compensator C added, $dP = 0$). Taking $\alpha_c = 0$ (for Invar), $\alpha_g L_g \approx 0$ for simplicity, (say L_g is very small)

$$(11.5 L_b - 23 L_a)\,10^{-6} = 0$$

$$L_b = 2L_a$$

therefore

$$L_c = L_b - L_a = L_a$$

Allowing for the gasket would require an increased length of compensator.

6.6 BOLTED ASSEMBLIES UNDER LOAD CHANGES
(Second year standard)

This section debates the force transfers in bolted assemblies under varying loads; the important aspect is to show that the highly stressed bolts see relatively little of the load fluctuations. The fluctuations are mostly absorbed by transfer of internal force between the components.

Consider the pressure container, figure 6.10a. To keep the notation consistent with the preceding section we refer to the attachments A, bolts B and gasket (if any) G. If there is no internal pressure the bolts are pretensioned to a force P which is fully reacted at the contact faces, figure 6.10a.

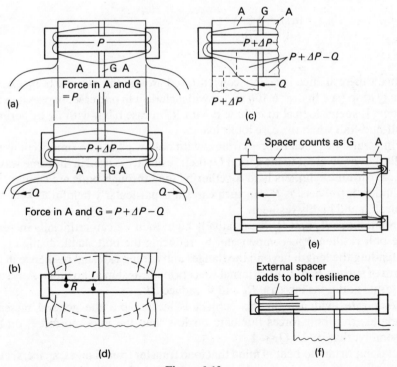

Figure 6.10

When the working load Q is applied, the bolts extend by a small amount x. The parts A and G also extend by x (if they did not, the contents would leak out), figure 6.10b.

The new bolt force $= P + \Delta P$. The contact force at the joint faces is now reduced by $Q - \Delta P$. By resiliences as defined in section 6.5

$$x = R_b \Delta P$$

and

$$x = (R_a + R_g)(Q - \Delta P)$$

for the bolts and the contacting parts respectively. Eliminating x gives

$$\Delta P = \frac{Q(R_a + R_g)}{R_a + R_b + R_g} \tag{6.22}$$

We are still left with the difficulty of assessing R_a. In a few cases there is a recognisable cross-section under uniform loading, for example, in castings where the bolt-hole may be surrounded by a local boss while the general walls are thin, or in cases of spacer-tubes. In the usual thick flanges the load is far from uniform. Some model tests carried out by the author show a roughly parabolic dispersion, figure 6.10d, with an initial slope roughly 0.6. This leads to the expression, per bolt

$$R_a \approx \frac{\ln\left(1 + \frac{1.2t}{R}\right)}{\pi E \sqrt{(R^2 - r^2)}} \tag{6.23}$$

Since this resilience is mainly concentrated around the nut and bolt seating, the Q loop (see figures 6.10b and c) will include it in one case, bypass it in the other. It seems logical to include it with R_g in case b but with R_b or perhaps half-and-half when the case looks like c.

In assemblies such as press frames with hollow pillars we simply count the pillars as equivalent to the gasket G; their resilience is used in the same way as R_g. Hydraulic cylinders held together by long through-bolts are similar; the cylinder barrel acts as R_g. In such cases it is particularly helpful to keep the bolt slim and highly resilient.

Sometimes R_a and R_g are relatively high; then we can artificially increase the bolt resilience to compensate, by reducing the bolt shank diameter, by extending the length beyond the flanges and adding external resilience in the form of springs or at least external spacers made as slim as possible. Any such external additions increase R_b and so reduce ΔP.

Accurate evidence on this subject is lacking but the general message remains, that bolt forces fluctuate far less than the imposed forces on the assembly, that is, $dP/Q \ll 1$.

It is important to bear in mind that load transfer from flanges, spacers, etc., can only continue as long as there is load in them, that is, as long as they are in compression. When the compression vanishes, so does the basis for the argument; the imposed force is then carried directly and totally by the bolts, tie-bars, etc. External spacers as in figure 6.10f continue to be in contact and therefore continue to add to the bolt extension.

Example 6.7

The cylinder head of a small engine is 60 mm deep and is secured by four bolts of 8 mm diameter, thread root diameter 6 mm. The lower 15 mm of the exposed bolt length are threaded (as in figure 6.8b). The effective cross-section of the head may be taken as 500 mm^2 per bolt. The head material has E = 7 × 10^{10} N/m^2, α = 23 × 10^{-6}/°C. The bolt material has E = 2 × 10^{11} N/m^2, α = 11.5 10^{-6}/°C. A 1 mm thick gasket of E = 8 × 10^9 N/m^2, α = 4 × 10^{-5}/°C is used; its area = 500 mm^2 per bolt. Find (a) the bolt stress increase due to a temperature rise, of 80 °C, (b) the further increase due to a gas force of 10 kN (2.5 kN per bolt) assuming the initial tightening to be adequate.

Working:

First find the resiliences, using N mm units

$$R_a = \frac{L}{AE} = \frac{60}{500 \times 7 \times 10^4} = 1.714 \times 10^{-6}$$

$$R_g = \frac{1}{500 \times 8 \times 10^3} = 0.25 \times 10^{-6}$$

R_b is taken in two parts, the plain part of the bolt and the threaded part, using say 6.5 mm as diameter. Resiliences in series are additive

$$R_b = \frac{\dfrac{46}{\pi \times 4^2} + \dfrac{15}{\pi \times 3.25^2}}{2 \times 10^5} = 6.84 \times 10^{-6}$$

$$\frac{dP}{dt} = \frac{-(61 \times 11.5 - 60 \times 23 - 1 \times 40)}{6.84 + 1.71 + 0.25} = \frac{718.5}{8.796} = 82 \text{ N/°C}$$

$$\text{Stress increase for 80 °C} = \frac{82 \times 80}{(\pi/4) \times 6^2} = 231 \text{ N/mm}^2 \text{ on root area}$$

$$\frac{dP}{dQ} = \frac{1.714 + 0.25}{1.714 + 0.25 + 6.84} = 0.223$$

$$\text{Stress for (dQ = 2500 N)} = \frac{0.223 \times 2500}{(\pi/4) \times 6^2} = 19.7 \text{ N/mm}^2$$

Example 6.8

A machine as in figure 6.10e has four columns 2.5 m long, 100 mm o/d, 60 mm i/d, end plates 200 mm thick and tie-bolts 55 mm o/d, with a thread of root diameter 49 mm. The tie-bolts are tightened to a stress of 300 N/mm² at the root. What load Q on the end plates will increase the stress to (a) 380 N/mm², (b) 500 N/mm²? E is the same for all the parts. Resilience of threaded part is negligible.

(a) Resiliences: neglect end plates in comparison with columns

$$R_g = \frac{2500}{\pi E (100^2 - 60^2)} \quad \text{all four}$$

$$R_b = \frac{2900}{\pi E \times 55^2} \quad \text{all four}$$

$$\frac{\Delta P}{Q} = \frac{R_g}{R_b + R_g} = \frac{2500}{6400 \left(\dfrac{2500}{6400} + \dfrac{2900}{3025} \right)} = 0.2895$$

$$\Delta P = 4(\pi/4) \times 49^2 \, (380 - 300) = 603\,437 \text{ N}$$

$$Q = 2.1 \text{ MN} \qquad \text{provided that columns are in contact}$$

$$\text{Tie-bolt force} = 4(\pi/4) \times 49^2 \times 380 = 2.86 \text{ MN}$$

$$\text{column force} = 2.86 - 2.1 = 0.76 \text{ MN}$$

(b) Same resilience but from previous answer we see that columns will be load-free hence the argument for load transfer vanishes

$$\text{Machine force} = \text{tie-bolt force} = 4(\pi/4) \times 49^2 \times 500 = 3.77 \text{ MN}$$

6.7 THE CATENARY AND SIMILAR PROBLEMS
(Advanced work)

This type of problem occurs in aerial ropeways, electricity transmission lines and stay-wires for masts. Figure 6.11 shows the basic equation and its design

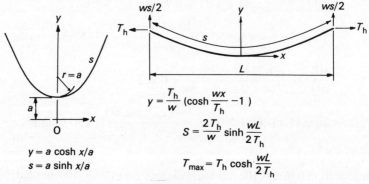

$$y = \frac{T_h}{w} \left(\cosh \frac{wx}{T_h} - 1 \right)$$

$$S = \frac{2T_h}{w} \sinh \frac{wL}{2T_h}$$

$$T_{max} = T_h \cosh \frac{wL}{2T_h}$$

$$y = a \cosh x/a$$
$$s = a \sinh x/a$$

Figure 6.11

implications. When $h \ll L$ these equations are greatly simplified by using the series expansions for sinh and cosh and setting $x = L/2$

$$h = \frac{WL^2}{8T_h} \qquad (6.24)$$

$$S = L\left(1 + \frac{W^2L^2}{24T^2}\right) \qquad (6.25a)$$

$$= L\left(1 + \frac{8h^2}{3L^2}\right) \qquad \text{via } 6.24 \qquad (6.25b)$$

$$T_{max} = T_h\left(1 + \frac{8h^2}{L^2}\right) \qquad (6.26)$$

Example 6.9

(1) Find the tensile stress in a steel wire rope spanning 2 km with a sag of (a) 100 m, (b) 40 m, assuming a density of 7800 kg/m³.

If cross-section = A, load per metre = $7800 \times 9.81 \times A$ N

(a) $T_h = \dfrac{7800 \times 9.81 \times A \times 4 \times 10^6}{800}$

or

(b) $\dfrac{7800 \times 9.81 \times A \times 4 \times 10^6}{320}$

Stress $= \dfrac{T_h(1 + 8h^2/L^2)}{A}$

Ans. 385, 957 MN/m².

The second value is close to breaking point for some steel wire ropes.

(2) Find the stresses if the rope is of 20 mm diameter and wind acting horizontally exerts a pressure of 1500 N/m² and if the temperature falls 30 °C.

Gravity loading per m = $7800 \times 9.81 \times \pi \times 0.01^2 = 24$ N

Wind loading per m = $1500 \times 0.02 = 30$ N

Resultant loading per m = $(24^2 + 30^2)^{1/2} = 38.4$ N

The tension and stress do not increase proportionately to the loading since any increase of sag affects the force balance equation 6.24, favourably.

Ropes are installed with tensioning screws, etc., so that we don't know the exact length. To attack the problem of how sag and extension behave we simply need to know what notional tension T_0 would exist in the rope if the sag were removed by external means. It does not matter whether T_0 is positive or negative since the notional state never happens.

Then we have the length $S = L$ when $T = T_0$.

When the rope is extended, the notional support being taken away, the tensile equation gives

$$\frac{S - L}{L} = \frac{(T - T_0)L}{AE}$$

If there is a temperature rise t, then for a coefficient of expansion α the length increases by an additional amount, so that

$$\frac{S - L}{L} = \frac{(T - T_0)L}{AE} + \alpha t L$$

The force equation 6.24 together with the catenary geometry of equation 6.25 and neglecting higher orders such as h^4/L^4 allow elimination of S and h, since $S/L = 1 + w^2L^2/24T^2$ (6.25a)

$$\frac{w^2L^2}{24T^2} = \frac{T - T_0}{AE} + \alpha t \tag{6.27}$$

Let us use this on example 6.9, part 2, sag 100 m. First, without temperature change. Condition 1, $T_1 = 385A$ N where A is in mm²

$$\frac{\left(\dfrac{24 \times 2000}{385A}\right)^2}{24} = \frac{385 - \sigma_0}{E}$$

$A = 100\,\pi$; E for ropes is about half the material value, due to the construction and effect of the soft centre built into many ropes, so we take 10^{11} N/m² or 10^5 N/mm²

$$4.56 \times 10^{-3} = 3.85 \times 10^{-3} - \sigma_0 \times 10^{-5} \quad \text{(strain)}$$
$$\text{(6.28, after 6.27)}$$

$$\frac{\left(\dfrac{38.4 \times 2000}{100\pi\sigma_2}\right)^2}{24} = \sigma_2 \times 10^{-5} - \sigma_0 \times 10^{-5}$$

Eliminating σ_0, from equation 6.28

$$\frac{2490}{\sigma_2^2} - \sigma_2 \times 10^{-5} = 0.71 \times 10^{-3} \tag{6.29}$$

For convenience, multiply by 10^5

$$\frac{2.49 \times 10^8}{\sigma_2^2} = 71 + \sigma_2$$

Trial procedure: σ_2/σ_1 should be less than 38.4/20, because of further sag (non-linear law). Try 640, 620, 600; $2.49\times10^8/\sigma^2 = 607, 648, 691$; $\sigma + 71 = 711, 691, 671$. Values cross near $\sigma = 607$; try 606, 607; 678, 675; 606 + 71 = 677; 607 + 71 = 678. True value just below 606 N/mm².

Second, with temperature change, 38.4 N/m load.
Using equations 6.29 and 6.27

$$\frac{2490}{\sigma_3{}^2} - \sigma_3\times10^{-5} = 0.71\times10^{-3} - 30\times1.1\times10^{-5}$$

$$\frac{2.49 \times 10^8}{\sigma_3{}^2} = 38 + \sigma_3$$

Splitting the difference, try $\sigma_3 = 620, 622, 624$; then $2.48\times10^{-8}/\sigma_3{}^2 = 648, 643, 639$; going the wrong way. Try 618; (651). Try 616; (656). By interpolation, $\sigma_3 = 616.5 \, \text{N/mm}^2$.

Comment:

(1) This is a long span; more load can be carried if increased sag is allowed.

(2) With higher initial tension, the effect of added load would be smaller.

The amount of ice build-up and wind loading on the iced-up cables, etc., varies according to country and climate and is laid down in codes of practice, in conjunction with permissible design stresses.

The other end of this problem area is a rope carrying a point load greatly exceeding the rope weight.

Figure 6.12 shows the principle on the left, followed by two practical forms:

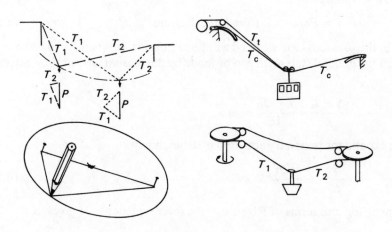

Figure 6.12

(a) a load free to ride on a carrying rope and pulled by a towing rope, (b) a load clamped to the rope, the whole rope being moved. Analytically these come to the same. In the first case the tension in the carrying rope is constant except for a contribution due to rope weight while the second case has two different tensions. For the same load and shape, clearly T_1 is equivalent to the sum of the forces $T_t + T_c$ while $T_2 = T_c$.

The shape of the locus of the load on a weightless inextensible rope would be an ellipse: one of the methods of drawing an ellipse uses the property that the distances from any point to the two foci add up to a constant. Taking a loop of string, we place pins at the foci and with a pencil keep the string taut while following round the curve. This does not quite agree with practice because we have ends that include tensioning devices and curved guides. The general case is well suited to graphical or computer-graphical solution. Here we confine ourselves to the horizontal, symmetrical case. When we come to ropes with significant self-weight relative to the point load, we shall return to the general aspect.

With the notation of figure 6.13

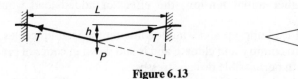

Figure 6.13

$$\text{length, } S = (L^2 + 4h^2)^{1/2} \qquad \text{from geometry} \qquad (6.30)$$

$T/\tfrac{1}{2}P = \tfrac{1}{2}S/h$ or

$$T = PS/4h \qquad \text{from force diagram} \qquad (6.31)$$

As in the previous case we need a notional tension T_0 when $P = 0$ and $S = L$. Then the general statement can be made by the tension law, strain = stress/E + αt

$$\frac{S - L}{L} = \frac{T - T_0}{AE} + \alpha t$$

We expand the geometry equation 6.30 binomially

$$\frac{S}{L} = 1 + \frac{2h^2}{L^2} \qquad (6.32)$$

Converting into terms of P by use of the force equation 6.31 gives, if $S \approx L$

$$\frac{S}{L} - 1 = \frac{P^2}{8T^2} \qquad (6.33)$$

We can now eliminate the unknown T_0 provided we have enough information on one or other of the states

$$T + AE\alpha t = \frac{AEP^2}{8T^2} + \text{constant} \tag{6.34}$$

This is again a cubic equation in T but is easily solved by tabulated trial-and-error means.

Example 6.10

An aerial ropeway carries one full skip of 300 kg per span of 50 m. Find the tension required for a sag of 0.8 m at the centre. Also find the tension and sag if the temperature is 30 °C below normal. The permissible stress at normal temperature is 250 N/mm², $E = 10^5$ N/mm² for this rope, $\alpha = 1.1\ 10^{-5}/°C$

(1) First condition: from equation 6.31

$$T = \frac{300 \times 9.81 \times 50}{4 \times 0.8} = 46\ 000\ \text{N}$$

A for stress of 250 N/mm² = 184 mm²

(2) At new temperature: use equation 6.34. First find constant using $t = 0$

$$T + 0 = \frac{184 \times 10^5 \times (300 \times 9.81)^2}{8 \times 46\ 000^2} + \text{constant}$$

$$\text{Constant} = 46\ 000 - 9414 = 36\ 600$$

$$T' - 184 \times 10^5 \times 1.1 \times 10^{-5} \times 30 = \frac{184 \times 10^5 \times (300 \times 9.81)^2}{8T'^2} + 36\ 600$$

$$T' - \frac{19.92 \times 10^{12}}{T'^2} = 6072 + 36\ 600 = 42\ 670$$

$$T' = 50\ 500\ \text{N} \qquad \sigma' = 274\ \text{N/mm}^2$$

Comment:

(1) Note that the increased stress is about three-quarters of the stress which the temperature drop would have produced in a fixed member while in the previous case it was much lower. This is not due to the point load versus the distributed load but due to the relatively lower sag, 1.6 per cent as against 5 per cent. A low sag implies high initial tension which makes the set-up more sensitive to temperature but less sensitive to load changes;

(2) The conventional sag for short-span aerial ropeways is 2 to 2.5 per cent. In mountain funiculars there may well be plenty of space below without incurring cost of tall pillars, so longer spans with deeper sag are possible. In

transmission lines the sag involves cost of high towers, to reduce the sag means shorter towers closer together. The solution adopted is optimised for capital cost, access difficulties, etc., while the cables are optimised for energy loss versus capital cost.

In many cases the point load and rope weight will be similar. The approach suggested for this is shown in figure 6.14. The catenary to either side of the point load neither knows nor cares how the forces upon it have been produced, so we pretend that the point load is an extra length of rope L_e such that $wL_e = P$, or $L_e = P/w$.

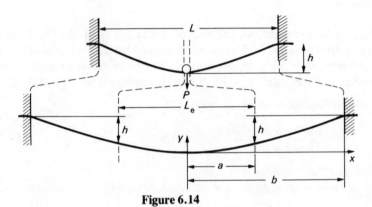

Figure 6.14

For convenience of writing we use values $a = \frac{1}{2}L_e$ and $b = \frac{1}{2}(L_e + L)$. Between $x = a$ and $x = b$ the catenary equation is as before, $y = T_h/w\,(\cosh wx/T_h - 1)$. If $h \ll L$ so that higher terms in the expansion are negligible, then

$$h = \frac{w(b^2 - a^2)}{2T_h} \tag{6.35}$$

Example 6.11

A load of 1529 kg is carried on steel rope of cross-section 653 mm² over a span of 400 m, sag 40 m. Find T_{max} and σ. Compare this with stresses for a point load and rope weight separately for the same sag. Take ρ as 7800 kg/m³

$$P = 1529g = 15\ 000\,\text{N} \qquad w = 653 \times 7800 \times 9.81 \times 10^{-6} = 50\,\text{N/m}$$

$$L_e = \frac{P}{w} = 300\,\text{m}$$

$$a = \frac{L_e}{2} = 150\,\text{m} \qquad b = \frac{L + L_e}{2} = 350\,\text{m}$$

We need the fundamental equation from figure 6.11: $y = (T_h/w)[\cosh (wx/T_h) - 1]$, expanded as the series

$$\cosh m = 1 + \frac{m^2}{2!} + \frac{m^4}{4!} \cdots$$

then using equation 6.35

$$h = y_b - y_a = \frac{T}{50}\left[\frac{50^2(b^2 - a^2)}{2T^2} + \frac{50^4(b^4 - a^4)}{24T^4} + \cdots\right]$$

Using the first term only to begin with

$$\frac{50(350^2 - 150^2)}{2T} = 40 \qquad T = \frac{50 \times 10^5}{80} = 62\,500 \text{ N}$$

Test for importance of second term: $(50\, b/T)^2/12 = 6.5 \times 10^{-3}$ – negligible

$$T = 62\,500 \text{ N} \qquad \sigma = 95.7 \text{ N/mm}^2$$

For load alone at same sag, $T = PL/4h$ from equation 6.25, $\sigma = 57.4$ N/mm^2.
For rope alone at same sag, $T = wL^2/8h$ from equation 6.18, $\sigma = 38$ N/mm^2.

For steep catenaries such as ropeways up mountain slopes, the approximations $h \ll L$, $L = S$ are no longer suitable. We return to the basic equations

$$y = \frac{T_h}{w}\left(\cosh \frac{wx}{T_h} - 1\right) \qquad \text{(see figure 6.8)}$$

$$\frac{dy}{dx} = \sinh \frac{wx}{T_h}$$

$$T = T_h\left[1 - \left(\frac{dy}{dx}\right)^2\right]^{1/2} \qquad \text{(triangle of forces at any point)}$$

$$T = T_h \cosh \frac{wx}{T_h} = T_h\left(\frac{wy}{T_h} + 1\right) = wy + T_h \qquad (6.36)$$

Figure 6.15 plots the basic generalised catenary $y/a = \cosh (x/a) - 1$. On the right is a useful scale showing the tension in terms of the horizontal rope length. This is the maximum tension at the upper end of a catenary of length x measured from the horizontal point.

For instance, if we require a rope to rise 50 m in a horizontal distance of 70 m with a sag of 8 m below the line-of-sight, we could use points $(0.3, 0.04)$ and $(1.0, 0.54)$. The top tension would be 1.55 times the rope weight whereas by a calculation as for a horizontal case the tension would be $wL^2/8h = 1.1$ times rope weight. The reason for the difference is that the bottom tension is pulling downwards or, in the limit, horizontally whereas in the horizontal solution both ends have an upward component of tension.

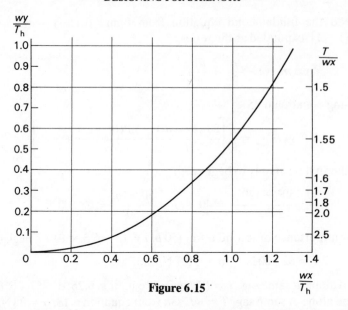

Figure 6.15

An important use of catenaries is in suspension bridges and overhead conductors for electric railways. In both, the loading consists of the curved catenary, the deck or lower conductor wire and the hangers. The interesting problems are due to flexibility. If a heavy load crosses a suspension bridge then at the supports the load path is stiff, the deck deflects very little. At other points the load is shared out and the deck is subject to bending, as shown exaggeratedly in figure 6.16. The effect can be reduced by using springs in the hangers nearest the towers, or sloping hangers. In railway conductors the same effect appears upside down. To maintain good contact the collector is pressed upwards with appreciable force, able to deflect the wires upwards at mid-span. At the supports, if rigid dropper wires are used, the effect is one of hard spots. With high-speed trains this would cause excessive wear. The remedies include flexible hangers, complicated catenaries and/or suspending the lower wire with such a curvature that when lifted by the collector each point comes up to a standard level. Another feature of railway conductors is the low headroom available, especially in Europe. To minimise changes of sag with temperature either the spans must be kept short or the tension is kept constant by means of weights and levers, etc.

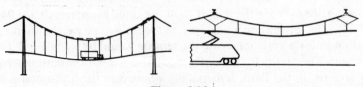

Figure 6.16

PROBLEMS

6.1 A chain as in figure 6.17a is 0.5 m long, attached to a drum of 0.1 m radius. If the mass of one link = 45 grams show that the effective density of the chain is about 11 200 kg/m^3 and find the angular velocity at which the stress at x–x reaches 80 N/mm^2 (that is, 80 MN/m^2). The material's density = 7800 kg/m^3.

Ans. 202 rad/s if gravity is neglected.
Repeat in the case of a chain 1.6 feet long, drum radius 4 inches, effective density 0.4 lb/in^3, that is, 691 lb/ft^3. $g = 32.17$ ft/s^2, stress = 11 600 lb/in^2. Note the inconsistent units.

Ans. 207 rad/s if gravity is neglected.

6.2 Find the tensile stress due to centrifugal action in the tyre of a road vehicle travelling at 600 km/h if the density of the tyre compound is 1100 kg/m^3.

Ans. 30.56 N/mm^2.

6.3 This question refers to section 6.3. A soft brass tube is tightly fitted into a thick steel block and sealed with soft solder. It is accidentally overheated to 350 °C; upon cooling it is found to leak. Suggest why. Can it be repaired?
Ans. The calculated stress is over 200 N/mm^2; brass probably yields in compression and shrinks appreciably before the solder sets again. Repair by re-soldering but avoid overheating.

6.4 A circular saw with fine teeth, of 0.25 m diameter with a 20 mm diameter hole runs at 3200 rpm. If the density = 7800 kg/m^3, $\nu = 0.28$, find the greatest tensile stress treating the disc as uniform.

Ans. 11.2 N/mm^2.

6.5 A joint is made of aluminium bronze, $\alpha = 18 \times 10^{-6}$/°C and uses stainless steel bolts, $\alpha = 16 \times 10^{-6}$/°C. The sizes and design are as in figure 6.17 except that Invar compensating spacers are to be let into the flange which will be counterbored to suit. Find the thickness (that is, length) of such spacers to fully compensate for thermal expansion but ignoring gasket expansion.

Ans. 10 mm.

6.6 A pressure vessel as in figure 6.17b has 36 bolts of 20 mm diameter body, thread root diameter 16 mm, mean effective thread diameter 17 mm for elastic purposes. The bolts are tightened to ¾ of yield point which is 640 N/mm^2; UTS = 800 N/mm^2, endurance limit ± 350 N/mm^2. The pressure cycles between 0 and 20 bar. Find the stress range in the bolts, draw a Smith diagram and estimate the fatigue life. Young's moduli are 190 000 N/mm^2 for

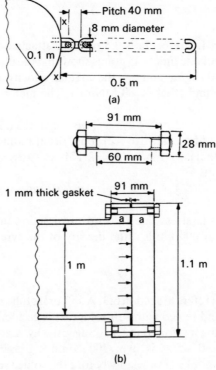

Figure 6.17

the bolts, 120 000 N/mm² for the apparatus and 16 000 N/mm² for the gasket. The bolt holes are drilled 21 mm diameter. Take the worst reasonable assumptions.

Ans. With stress concentration, highest stress levels off at yield point, range: 640 at 20 bar, 470 at 0 bar, approximately. Life indefinite.

6.7 A telephone cable of 1.1 kg mass per metre is supported by a steel catenary wire of 25 mm² cross-section, weighing 0.22 kg/m between supports 50 m apart. Find the sag if the stress as installed is to be 300 N/mm² in the catenary, zero in the cable.

Ans. 0.54 m.

6.8 In the installation of problem 6.6 find the stress due to an additional load of 1 kg/m, for example, due to ice, but with no temperature change.

If the cable remains slack, *ans.* 366 N/mm².

6.9 What would the stress become in the case of problem 6.8 if the temperature falls 40 °C and $\alpha = 11 \times 10^{-6}/°C$?

If the cable still remains slack, *ans.* 423 N/mm².

7 BEAM PROBLEMS

Beam problems occur in practically all engineering designs, not always in obvious form. The designer's chief interest lies in finding the highest stresses, deflection at the load points and very occasionally deflections elsewhere. Many students are taught to use Macaulay's method and the three moment equation. These excellent methods have the following disadvantages: they do not aid a sense of structure, they are lengthy, errors are difficult to spot, they are not very amenable to assessing the effect of non-rigid end fixings, they become exceedingly lengthy in beams of varying section which discourages any attempts to optimise the beam in accordance with the bending moment at various points.

The methods adopted here are the Myosotis method (superposition of basic cases), moment area for slope and for finding the moment distribution, and strain energy of standard cases applied section by section.

7.1 BASICS

For our purposes a beam is any member in bending; straight or curved, uniform or of varying section. We shall neglect axial forces unless they have a moment about the centroid and we shall neglect shear deflections in the analysis.

The simpler cases are statically determinate; we can find the support forces directly from the external loads or moments and sketch the bending moment diagram, that is, the plot of the bending moment along the beam. Other cases are statically indeterminate, for example, beams built-in at the ends, beams on three or more supports, beams on resilient foundations; in these types there is interaction between the beam deflection and the support parameters.

For convenience we take the x axis along the beam and use the y axis for deflection normal to the beam. This is of course inconsistent with chapter 2 where y was used as an internal dimension within the beam; therefore we shall not consider stresses within the beam in this chapter. If stresses are referred to they will generally be the maximum stress due to bending, $\sigma = M/Z$ (see section 2.2.1). In curved beams and portal frames suitable conventions will be specified locally.

If a beam is substantially straight, the bending law $M/I = E/R$ can be simplified. In any plane curve, the radius of curvature

$$R = \frac{\left[1 + \left(\dfrac{dy}{dx}\right)^2\right]^{1\frac{1}{2}}}{\dfrac{d^2y}{dx^2}}$$

When $dy/dx \ll 1$ this simplifies to $1/R = d^2y/dx^2$; hence (ignoring sign convention for now)

$$\frac{d^2y}{dx^2} = \frac{M}{EI} \tag{7.1}$$

Integrating gives the slope equation

$$\theta = \frac{dy}{dx} = \int \frac{M}{EI}\, dx \qquad \text{with suitable limits} \tag{7.2}$$

The next integration gives the deflection

$$y = \int \theta\, dx \tag{7.3}$$

In some cases we need to go the other way, differentiating the moment equation

$$\frac{d^3y}{dx^3} = \frac{d}{dx}\left(\frac{M}{EI}\right) \tag{7.4}$$

The most used application of this equation is when E and I are both constant, which is by far the commonest situation anyway. Remembering section 2.2, we see that $dM/dx = P$. As discussed briefly then, P is not necessarily just one force, it is the net resultant of all transverse (that is, vertical) forces to one side of the point under consideration, in other words the net shear force at that section

$$\left|\frac{d^3y}{dx^3}\right| = \frac{P}{EI} \qquad \text{(if } EI = \text{constant)} \tag{7.5}$$

$$\text{or} \qquad \frac{P}{EI} - \frac{M\left(\dfrac{dI}{dx}\right)}{EI^2} \qquad \text{if } I \text{ also varies}$$

On some occasions we need the next differential

$$\left|\frac{d^4y}{dx^4}\right| = \frac{d}{dx}\left(\frac{P}{EI}\right) \tag{7.6}$$

If E and I are constant

$$\left|\frac{d^4y}{dx^4}\right| = \frac{w}{EI} \qquad \text{where } w \text{ is the load per unit length} \tag{7.7}$$

Equation 7.5 is rarely used but 7.7 is used in sections 7.10 and 9.5. See also appendix A.8.

The sign convention adopted in this chapter is a compromise between logic and convenience. Downward loads are treated as positive but deflection y is negative if below the line of the unloaded beam. The advantage of this is that slope and curvature are then as in mathematics, positive θ is an up-slope, positive d^2y/dx^2 is a minimum.

In sections 7.10 and 9.5 and appendix A.8 it is necessary to adopt a fully consistent rule to keep the differential equation self-consistent, with w and y of the same sign so that $d^4y/dx^4 \propto y$.

7.2 MOMENT AREA AND ENERGY METHODS

In many instances the methods about to be described give answers much more quickly and with less risk of arithmetical error than the full integration procedure.

7.2.1 Moment Area

This method makes direct use of equation 7.2 where E and I are constant. Taking advantage of the fact that the bending moment diagram is usually obvious by inspection, being a triangle, trapezium or parabola, the integral $\int M \, dx$ is also immediately apparent between any two convenient points. The slope change is simply area of moment diagram/(EI).

As an example, consider a simply supported beam of span L with a central load P, figure 7.1. With the centre as starting point, the area of the moment diagram up to the end is

$$\tfrac{1}{2} \text{ base} \times \text{height} = \tfrac{1}{2} \times \frac{P}{2} \times \frac{L}{2} \times \frac{L}{2} = \frac{PL^2}{16}$$

The centre slope $= 0$ by symmetry, therefore the end slope $= PL^2/16EI$.

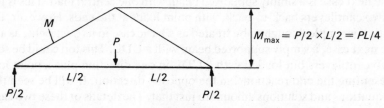

$$M_{max} = P/2 \times L/2 = PL/4$$

Figure 7.1

7.2.2 Energy of Bending

This method often provides the quickest way to the deflection under a point load. For multiple or distributed loads it is difficult and can go astray.

In an elastic loading build-up the deflection is proportional to the force or moment causing it (except in buckling-type cases, etc.). The average force is therefore half the final value, and the same for moments. In bending, if a moment M results in an angle change θ, the energy $U = \frac{1}{2}M\theta$.

From equation 7.2, $\theta = \int M/EI \, dx$, therefore

$$U = \int \frac{M^2}{2EI} \, dx \tag{7.8}$$

The same argument goes for a point load P, hence $U = P\Delta/2$ if Δ is the final deflection. In a structure subjected to a point load P, the bending deflection is given by

$$P\Delta = \int \frac{M^2}{EI} \, dx \tag{7.9}$$

and there is no need for I to be constant; E also can vary (this is rather unusual).

7.3 STANDARD CASES

Many practical problems are standard cases or can be derived from standard cases. Figures 7.2 and 7.3 show the most common ones. Note the general expressions, the maximum values and the strain energy.

The type shown in figure 7.2 is a cantilever beam or cantilever for short. It is somewhat abstract since a truly built-in end is not realistic, there is invariably some flexure inside the fixing and in the surroundings. A true zero end slope occurs by symmetry when two equal loading systems from opposite sides meet. P stands for a point load; this also is unrealistic, a localised load is distributed over a short distance. The symbol used for a distributed loading is w per unit length. Unless otherwise stated, w implies a uniform loading, UDL for short.

The next case is a simply supported beam with one central load. This is just like two cantilevers back-to-back, with point loads at the ends. However, this situation is simple enough to be treated as a basic case in its own right, as also is the next case, a simply supported beam with a UDL. This too could be seen as two cantilevers but loaded with a UDL in one direction plus a point load representing the end reactions, in the opposite direction. It will be seen that the equations and solutions amount to just that. The details of these two cases are presented in figure 7.3.

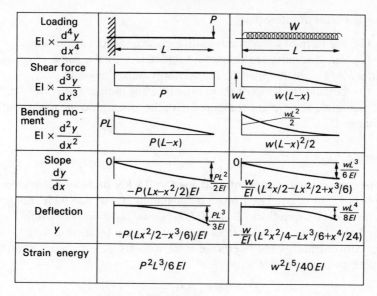

Figure 7.2

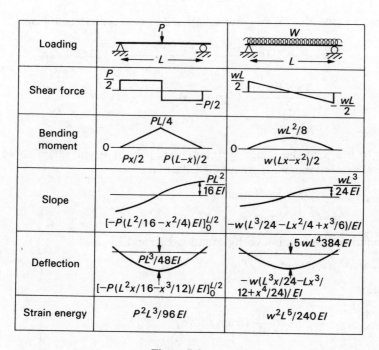

Figure 7.3

Some points of interest are

(1) the form of support shows that spanwise forces are absent;

(2) the diagrams are drawn approximately to the same scale of total load, $wL = P$, which shows the effect of spreading the load;

(3) beyond the support there is no bending moment provided the beam is of negligible self-weight compared with the loading, so the slope continues unchanged.

Example 7.1

An I-beam carries a concentrated load of 15 kN midway between simple supports 1.2 m apart. The I-beam is 200 mm high, 100 mm wide, flange 10.5 mm thick, web 6 mm thick. $E = 200\,000\,N/mm^2$, $G = 78\,000\,N/mm^2$.

Find the highest tensile stress and shear stress (section 2.4.3), ignoring stress concentrations and fillet radii at the web–flange junction. Estimate the bending and shear deflections.

$$I = \frac{(100 \times 200^3 - 94 \times 179^3)}{12} = 21.74 \times 10^6$$

$$\sigma = \frac{My}{I} = \frac{15 \times 10^3 \times 1.2 \times 10^3 \times 100}{4 \times 21.74 \times 10^6} = 20.7\,N/mm^2$$

To find τ at the neutral axis, use equation 2.27 over limits 0 to 100 mm

$$\int by\,dy = 100^2 \times \frac{100}{2} - 89.5\,\frac{(100 - 6)}{2} = 123\,500\,mm^3$$

$$P = \frac{15 \times 10^3}{2} \qquad \text{(figure 7.3)}$$

$$\tau = 7.1\,N/mm^2$$

Now to find τ at the flange–web junction, $\int by\,dy$ is still taken about the neutral axis but the limits are now 89, 100 mm

$$\int by\,dy = 99\,500\,mm^3$$

$$\tau = 5.72\,N/mm^2$$

Maximum stress at this level by equation 2.33 and 2.7

$$y = 89\,mm$$

therefore

$$\sigma = 20.7 \times \frac{89}{100}\,\text{(from above)}$$

$$= 18.4\,N/mm^2$$

$$\tau_{max} = \left(\frac{\sigma^2}{4} + \tau^2\right)^{1/2} = 11.0 \, N/mm^2$$

Bending deflection $= \dfrac{PL^3}{48EI} = \dfrac{15 \times 10^3 \times 1.2^3 \times 10^9}{48 \times 2 \times 10^5 \times 21.7 \times 10^6}$

$$= 0.124 \, mm$$

Shear deflection using average of the web shear stresses found above, that is, $(7.1 + 5.72)/2$

$$\text{Deflection} = \frac{6.41 \times 600}{78\,000} = 0.05 \, mm$$

Comment: shear deflection is appreciable because the beam span is unusually short in relation to depth. The governing stress for yield is at the web–flange junction but is only slightly above the shear stress at the outer surface, $\sigma_{max}/2$. The stress also is slightly exaggerated since a 15 kN load must occupy a finite piece of span, otherwise the web may yield in compression under the load (see chapter 8).

7.4 SOME APPLICATIONS

7.4.1 Simple Cases

Now we take a few examples of applying the basic cases to various loading conditions. The overhung loads, figures 7.4a and b are treated by first calculating the reaction R. Figure 7.2 gives the deflection of the cantilever AB. Offsetting this upwards gives A'; from this we draw A'BC'. Then C'C is the deflection of cantilever BC. Once we understand the shapes we can proceed either graphically or by calculation. The deflections are taken directly from the standard cases. This method also makes it easy to allow for deflections at the supports; such deflections simply increase the slope of A'C'. If only the deflection at P is required, energy is the quickest way. $P\Delta/2 = (R^2a^3 + P^2b^3)/6EI$, $\Delta = P(ab^2 + b^3)/3EI$.

For the case in figure 7.4c we prefer to set the tangent at P horizontally on the paper. Then taking $R_1 = Pb/(a + b)$, $R_2 = Pa/(a + b)$ gives the deflections of each end relative to the tangent. The line between them is the true horizontal. Drawing this on graph paper to an extended vertical scale gives the deflections at P and elsewhere, using the values from figure 7.2. If we only require the deflection at P it is easier to use energy. The strain energy $= R_1^2a^3/6EI + R_2^2b^3/6EI$; this must equal $\frac{1}{2}P\Delta_p$. Substituting for the reactions $R_1 = Pb/(a + b)$ $R_2 = Pa/(a + b)$

$$\Delta_p = \frac{Pa^2b^2}{3(a + b)EI} \tag{7.10}$$

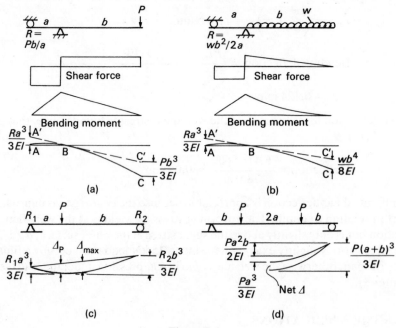

Figure 7.4

The two symmetrical loads at (d) can be broken down into cantilevers since by symmetry the slope at the centre = 0. The end reaction which in this case = P gives an upward deflection $P(a + b)^3/3EI$. The actual load gives a downward deflection = $Pa^3/3EI$; beyond this, part b extends at the slope of $Pa^2/2EI$ and thus rises by a further $Pa^2b/2EI$. This gives the net deflection at the centre as $P[(a + b)^3/3 - a^2(a/3 + b/2)]/EI$.

The deflection at the load points is obtained more easily by energy. The work done by the *two* loads $P = P\Delta_p$. The strain energy of the two end sections = $2P^2b^3/6EI$, while the centre section, length $2a$, constant bending moment Pb has by equation 7.5 strain energy of $P^2b^2 \times 2a/2EI$. Thus

$$P\Delta_p = \frac{P^2\left(ab^2 + \dfrac{b^3}{3}\right)}{EI} \quad \text{or } \Delta_p = \frac{P\left(ab^2 + \dfrac{b^2}{3}\right)}{EI}$$

7.4.2 Advanced Cases (Second year work)

Some beams are reinforced locally so that I is not uniform. The energy method is very convenient here. The basic beam has an I value I_1, with reinforcement to bring it up to I_2 near the loading point. The bending moments are not affected by the reinforcement, so the energy items can be written down directly, as seen in figure 7.5a. Section *a* is obvious, the energy item is taken directly from figure 7.2. Section *b* is treated by difference. First we consider the whole length *a* + *b* as a standard case, then we take away the

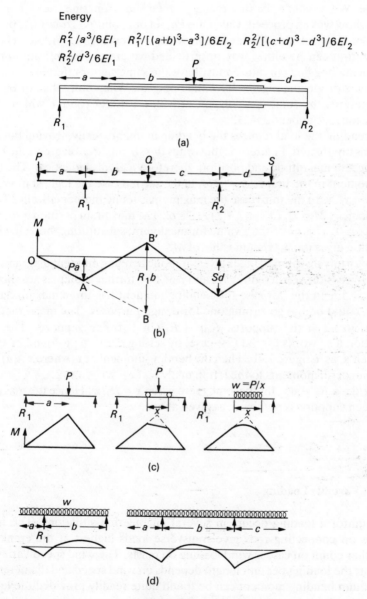

Figure 7.5

energy which section a would have contained if it had been of I value I_2. Sections d and c are treated like a and b. The total energy is added up and made equal to $P\Delta/2$.

For design purposes generally, and for the placing of reinforcement, we need some skill in finding the magnitudes and positions of high bending moments, by plotting bending moment diagrams. With point loads this is very simple. We calculate the first reaction, then start from one end. Figure 7.5b shows how we can proceed. Calculate Pa, set out point A. Draw OA, produce to meet the next load line at B. Draw BB' to represent R_1b, etc. The sign convention can be either way up provided we are consistent; all moments producing hogging curvature must have the same sign, preferably minus, all moments producing sagging have the opposite. When more than three loads are present, the funicular polygon provides a useful routine which avoids confusion, see appendix A.12.

Spreading the load reduces the bending moment locally, leaving the other regions unaffected. Figure 7.5c illustrates this. In all three diagrams the load is identical in magnitude and position, therefore R_1 is also identical. The bending moment in the first case $= R_1a$, in the second case the highest moment is $R_1 (a - x/2)$. In the third case the maximum is found mathematically. Taking a variable z, $M = R_1 (a - x/2 + z) - wz^2/2$. The maximum position is found by $dM/dz = 0$. This gives the z value for maximum; substituting this value in the equation gives the maximum value of M.

Uniformly distributed loads give parabolic bending moment diagrams. The basic forms are seen in figures 7.2 and 7.3. Two further examples are shown in figure 7.5d. In the last case the bending moment has three maxima, any of which could be the governing one for design purposes. The magnitudes are $wa^2/2$, $wc^2/2$ at the supports, $R_1a - R_1^2/2w$ between supports. The same equation also works for the first case, by setting $c = 0$. R_1 is found by taking moments, as in figure 7.4b. Then the bending moment at x (where $x > a$) is the net sum of all moments to the left, namely $R_1 (x - a) - wx^2/2$.

$dM/dx = R_1 - wx$; this becomes zero when $x = R_1/w$. Hence the maximum between supports is obtained, as given above.

7.4.3 Variable Loading

Non-uniform loading occurs in several engineering situations. The inertia forces on connecting rods give transverse loads highest at the crank end. Another common case is wind loading on masts. The wind speed varies with height; the loading per unit length depends on wind speed and diameter. The maximum bending moment can be found quite readily provided the loading can be expressed in a form suitable for integration, preferably a polynomial.

Figure 7.6 shows a linearly varying loading on a simply supported beam. The treatment is readily extended to non-linear loadings.

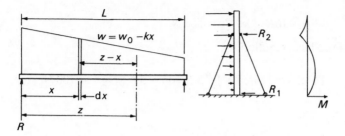

Figure 7.6

To find the reaction R, take moments about the other end.

$$RL = \int_0^L (w_0 - kx)(L - x)\mathrm{d}x = \frac{w_0 L^2}{2} - \frac{kL^3}{6} \tag{7.11}$$

The bending moment is due to all forces to one side of the point required. As we do not know which point has the maximum, we use a temporary variable z

$$M = zR - \int_0^z (w_0 - kx)(z - x)\mathrm{d}x = zR - \left(w_0 z^2 - \frac{kz^3}{2} - \frac{w_0 z^2}{2} + \frac{kz^3}{3} \right)$$

$$M = zR - \frac{w_0 z^2}{2} + \frac{kz^3}{6} \tag{7.12}$$

To find the greatest bending moment we differentiate with respect to z, set to zero obtaining a quadratic in z. The root is substituted into equation 7.12.

If the load extends beyond the supports by an amount b, equation 7.11 will have different limits and equation 7.12 will have $M = (z - b)R$, etc.

If the beam is of non-uniform section, then the point of maximum bending moment is not the point of maximum stress. It is generally quite complicated to find the latter directly; a simpler way is to make two or three trial calculations near the point of maximum moment, towards the slimmer side.

Example 7.2

A chimney is subjected to wind force which increases with height such that the force per metre $= 3.5h$ N at a height h (because wind speed varies as $\sqrt{h}$ and force varies as speed^2). The chimney is 20 m high and is supported by ropes at 16 m height. Assuming this support and the base to be simple supports not exerting any couples, find the greatest bending moment (see figure 7.6).

(1) Take moments about the base

$$\text{Wind moment} = \int_0^{20} 3.5h \times h \times \mathrm{d}h = 3.5 \times \frac{20^3}{3} = 9333 \text{ N m}$$

$$= 16R_2 \qquad R_2 = 583 \text{ N}$$

(2) Find base reaction R_1:

$$\text{Total wind force} = \int_0^{20} 3.5\,h\,\mathrm{d}h = 3.5 \times \frac{20^2}{2} = 700 \text{ N}$$

By force balance

$$R_1 = 700 - R_2 = 117 \text{ N}$$

(3) Bending moment at R_2

$$\mathrm{d}F = 3.5h\,\mathrm{d}h \qquad \text{distance} = (h - 16) \text{ m}$$

$$M = \int_{16}^{20} 3.5h\,(h = 16)\,\mathrm{d}h = 3.5\left[\frac{h^3}{3} - \frac{16h^2}{2}\right]_{16}^{20} = 4555 - 4032$$

$$= 523 \text{ N m}$$

(4) Bending moment at z, between R_1 and R_2

$$M = R_1 z - \int_0^z 3.5h(z - h)\,\mathrm{d}h$$

$$= 117z - \left[\frac{3.5h^2z}{2} - \frac{3.5h^3}{3}\right]_0^z = 117z - \frac{3.5\,z^3}{6}$$

(5) To find maximum, differentiate with respect to z and set to zero

$$117 = 3.5 \times \frac{3z^2}{6} \qquad z^2 = \frac{117 \times 6}{3 \times 3.5} \qquad z = 8.176 \text{ m}$$

Substitute in (4); maximum moment between R_1 and $R_2 = 957 - 318 = 639$ N m

(6) Answer: 639 N m at 8.2 metres. Upper bending moment smaller.

(7) Comment: as in practice there would be a slight base moment, this seems a good choice of support height.

7.5 STATICALLY INDETERMINATE BEAMS

In many instances beams are fixed or partly fixed at the supports. Then the forces and moments are not determined by the load alone. The first pair of cases are built-in beams. In practice these would occur as part of a long row so that by symmetry the ends cannot have a slope. If literally built in there is some deflection in the fixing and the restraint is incomplete. The ideal cases are shown in figure 7.7 with a central point load in (a), a UDL in (b). The solid lines show the bending moment diagram due to the load alone. If the ends cannot change slope then there must be a restraining moment. By the moment area method (section 7.2) the bending moment diagram must have an area of zero. This tells us the magnitude of the restraining moment, shown dashed. The restraining moment is constant right across when acting alone; if it has the same area as the diagram shown solid, then it is of the right magnitude, leaving the resulting net moment diagram (shaded). At (a) it is half the original height, by inspection. At (b) it is 2/3 of maximum (see section 2.4.3, figure 2.27f), since the curve is a parabola.

To find the deflection at the centre we first find the deflection due to a pure moment and then subtract it from the standard case.

Figure 7.7c shows the circular shape of a beam under a constant moment M. From the bending laws, $R = EI/M$. By the theorem of crossed chords, $y(2R - y) = (L/2)^2$; if $y \ll L$, $y = L^2/8R = ML^2/8EI$.

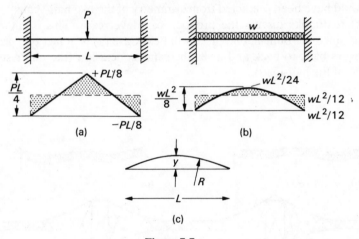

(a)

(b)

(c)

Figure 7.7

In case (a) the final deflection is the simply-supported deflection from figure 7.3 minus the circular arc deflection due to the end moment $M = PL/8$. This gives

$$y = PL^3/48EI - \frac{PL}{8} \times \frac{L^2}{8EI} \quad \text{from above}$$

so that

$$y = \frac{PL^3}{196EI} \tag{7.13}$$

In case (b) the end moment $= wL^2/12$, giving

$$y = \frac{5wL^4}{384EI} - \frac{wL^2}{12} \times \frac{L^2}{8EI} = \frac{wL^4}{384EI} \qquad (7.14)$$

This prompts the question of how much upward force would be needed to raise the middle up to the load-free level. In the point-loaded beam the answer is obvious, it is just P. In the UDL cases we find the required force Q by using the standard cases of deflection. Figure 7.8a, using figure 7.3, $QL^3/48 = 5wL^4/384$

$$Q_a = \frac{5}{8} wL \qquad (7.15)$$

In (b), using equations 7.13 and 7.14: $QL^3/192 = wL^4/384$

$$Q_b = \frac{1}{2} wL \qquad (7.16)$$

This could have been predicted from symmetry of the two half-spans.

Most textbooks discuss the propped cantilever case. This is rather unrealistic since the built-in end piece will flex. Case (a) is in fact two propped cantilevers back-to-back and gives the same conclusion (the outer support takes $^3/_8$ of the half-load).

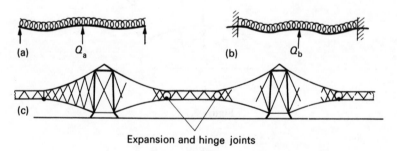

Expansion and hinge joints

Figure 7.8

The disadvantage of built-in beams is that if the supports move because of ground settlement the beams may develop high stresses. This can be overcome by working out the position of the points of zero bending moment (points of inflection) and constructing these as hinges which transmit shear force but give way in bending. The Forth railway bridge is built in this way, as sketched in figure 7.8c.

7.6 OPTIMUM SUPPORT POINTS

This topic is important in the design of storage hoppers, retaining walls, gantry cranes, etc., where there is bending due either to a distributed loading or due to a rolling load. We may well have choice over the placement of supporting frames, ribs or pillars. The main beam or plate is usually of equal strength in either direction and we confine ourselves to this kind. If it is unsymmetrical in strength the principles are the same but the execution becomes more elaborate. The principle is to place the reaction points such that the bending stresses reach the permitted maximum in as many situations as possible; then we are making the fullest use of the material properties; we are using the minimum of material and also obtain more or less the lowest deflections.

To begin with we consider a UDL over two supports. Figure 7.9 shows the loading, shear force and bending moment diagrams for a given span and loading, to scale, with varying support spacing. Case (c) is nearly the best, permitting the slimmest beam for a given duty, since the beam size required depends on the greatest moment.

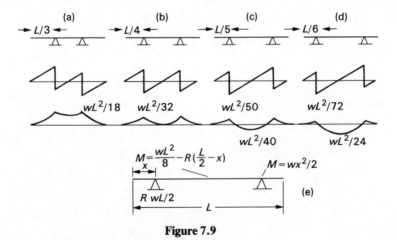

Figure 7.9

To find the best position analytically we use the notation shown in (e). The best solution will be that which makes the moment at the support numerically equal to the central moment and of opposite sign $wx^2/2 = -wL^2/8 + wL(L/2 - x)/2$. This equation when simplified will be found to have one positive root, $x = (1/\sqrt{2} - \frac{1}{2})L = 0.207L$. $M = 0.0214wL^2$.

Notice how the bending moment has a maximum or minimum wherever the shear force is 0, even when at a discontinuity.

If we have three symmetrical supports on a common level then there are three places of high bending moment: over the outer support, in the centre or in between. Figure 7.10 shows how the bending moments are distributed. We cannot expect to make all three maxima equal. The analysis uses standard cantilever cases from figure 7.2, with the condition that at the outer support the deflection due to the reaction equals that due to the UDL. The centre is fixed (by symmetry); thus we obtain the magnitude of the reaction for any position and from this the bending moments. The best position for the outer supports is 29 per cent of the half-span from the end; this makes the central and outer bending moments equal. The in-between moment is smaller.

The analysis is relatively advanced. First we find the deflection at x due to a UDL of magnitude $w = 1$. If the half-span $a = 1$

$$\text{the deflection at } x = \frac{\left(\dfrac{x^2}{4} - \dfrac{x^3}{6} + \dfrac{x^4}{24}\right)}{EI} \qquad \text{(from figure 7.2)}$$

The deflection due to an upward force R is $Rx^3/3EI$. Since these must be equal if the supports are level

$$R = 0.75/x - 0.5 + 0.125x$$

The bending moment at the centre $= wa^2/2 - Rx$

$$= 0.5 - 0.75 + 0.5x - 0.125x^2$$

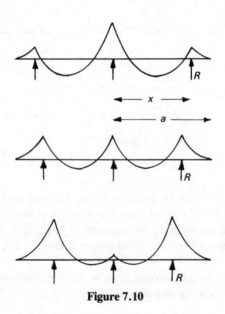

Figure 7.10

For optimum x this moment is made equal to the overhang moment $w(a - x)^2/2 = (1 - x)^2/2$. This simplifies $^5/8x^2 - 1\frac{1}{2}x + \frac{3}{4} = 0$; $x = 0.71$. The bending moment between supports is lower than the higher of these.

A rolling load has two critical positions – at the end beyond the support and between supports. With two hinged supports the case is very simple; figure 7.11a shows the bending moment diagrams. To equalise the bending moments, $Px = P(L - 2x)/4$; $x = L/6$. If the supports are rigid, $Px = P(L - 2x)/8$ (from figure 7.3), $x = L/10$ for the lightest beam. The bending moments are shown in (b). With three *hinged* supports (c) the lightest beam design results if we make x about $L/12$, while with three *rigid* supports the case is identical with (b). Truly rigid supports are rather improbable in practice.

The sharp peaks are theoretical; in practice the so-called point loads are distributed over a certain distance, rounding off the diagrams and reducing the peak values.

The regions of low bending moment are of interest when beams have to be joined, because of limited stock length, transport and assembly problems, etc. A joint is generally weaker than an unbroken section so we place the joints away from the peaks.

These principles can be seen put into practice in cranes for container depots, railway goods yards, etc.

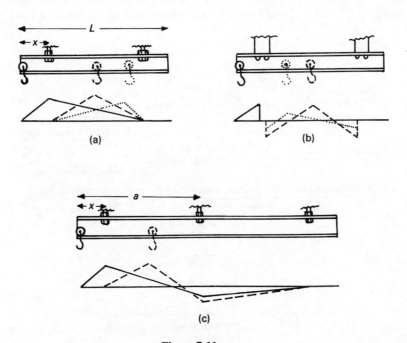

(a)

(b)

(c)

Figure 7.11

7.7 OPTIMISING BEAM PROFILES
(This and the rest of chapter 7 is second to third year work)

In addition to the best placing of supports we only need to look at figure 7.5 and more particularly at figure 7.7 to realise that there must be many savings possible by reducing the beam section where the bending moments are small. When using uniform material such as timber joists, rolled I-beams, etc., the extra cost is only rarely justified but when cast metals, concrete or large welded fabrications are concerned the extra trouble can be well worth while. Examples from the past are eighteenth century rails of edge and'plate' type, nineteenth century plate girder bridges (riveted rather than welded) and dry-dock gates (figure 7.12a to d). In all these cases the depth is reduced where the bending moment is smaller. It is little used in smaller-scale work largely because the analysis seems too laborious and is not generally given in textbooks. Full optimisation to constant stress (that is, equal peak stresses at all cross-sections) is hardly practical because of the points of inflexion (zero bending moment) where only shear is present. Below is presented a simplified version of an optimised beam (e), taken from one of the author's design studies for use in the heavy engineering industry. The beam is considered built-in at A and B, by symmetry.

The non-uniform cross-section means that the standard case from figure 7.3 cannot be used. To find the bending moment distribution we use the basic equation 7.2, $\Delta\theta = \int M/EI \, dx$. From symmetry we know the beam is hori-

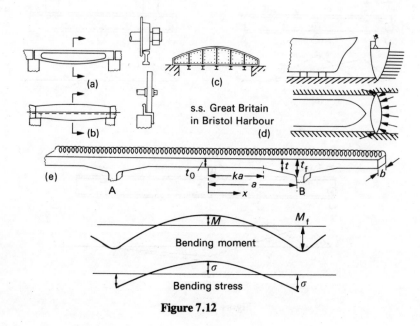

Figure 7.12

zontal at $x = 0$ and again at $x =$ a. Hence the integral between these limits must equal 0.

We take this in two stages. From 0 to ka the I value is constant, $bt_0^3/12$. We assume a bending moment M_c at the centre, from symmetry, shear force $= 0$ at the centre. Thus the first part is

$$\int_0^{ka} \left(M_c - \frac{wx^2}{2} \right) \times \frac{12}{Ebt_0^3} \, dx$$

For the second part we have varying t in the denominator. We tackle this by first defining t in terms of x: $t = m + nx$, therefore

$$x = \frac{t - m}{n} \tag{7.17}$$

Then we change variables from x to t

$$dx = \frac{dt}{n} \tag{7.18}$$

$$\int_{ka}^{a} \left(M_c - \frac{wx^2}{2} \right) \times \frac{12}{Ebt^3} \, dx$$

converts to

$$\int_{t_0}^{t_f} \left[M_c - \frac{w(t - m)^2}{2n^2} \right] \times \frac{12}{Ebt^3} \frac{dt}{n} \tag{7.19}$$

This can be integrated but much labour is saved by making $m = 0$. This is also quite beneficial structurally as we shall see shortly.

The physical meaning of n is the slope of the beam outline. By getting rid of m we have started the slope at the origin, thus $n = t_f/a$.

Since we shall set the total integral to zero, we can cancel common constants such as $12/Eb$. Integrating both parts

$$\left[\frac{M_c x}{t_0^3} - \frac{wx^3}{6t_0^3} \right]_0^{0.6a} + \left[\frac{-M_c}{2t^2 n} + \frac{w \ln t}{2n^3} \right]_{t_0}^{t_f} = 0$$

Substituting $t_0 = 0.6t_f$ and putting in the limits

$$\frac{0.6aM_c}{(0.6t_f)^3} - \frac{w(0.6a)^3}{6(0.6t_f)^3} - \frac{aM_c[1/t_f^2 - 1/(0.6t_f)^2]}{2t_f} + \frac{wa^3 \ln 0.6}{2t_f^2} = 0$$

$$2.778aM_c - 0.1667wa^3 + 0.333aM_c - 0.2554wa^3 = 0$$

$$M_c = 0.133wa^2$$

To find M_f use $x = a$; this gives

$$M_f = \frac{wa^2}{2} - 0.133wa^2 = 0.367wa^2$$

The stress at the centre $= \dfrac{6M_0}{t_0^2} = \dfrac{6 \times 0.133wa^2}{(0.6t)^2} = 2.217\dfrac{wa^2}{t^2}$

if $b = 1$

At the end

$$\text{stress} = \frac{6M_f}{t^2} = \frac{6 \times 0.377wa^2}{t^2} = \frac{2.262wa^2}{t^2}$$

This confirms the good choice of $k = 0.6$, (arrived at by trial and error).

There are of course other variants; they may save more material but can be more lengthy to solve, leading to risk of errors.

Just for interest, the point of inflexion is found where $M_c - wx^2/2 = 0$, $x = 0.515a$; we should therefore check for stress at $x = 0.6a$, $0.8a$, etc.

$$M_{0.6} = 0.133 - \frac{0.6^2}{2} = -0.047 \quad \text{no problem}$$

$$M_{0.8} = 0.133 - 0.32 = -0.187$$

stress check gives $0.187/0.8^2 = 0.292$, smaller than 0.367.

7.8 CURVED FRAMES

These are of limited importance in practice but form a good introduction to the widely used portal frames, etc. One use of such frames is in proving rings, a simple form of load cell. The stiffness of such a ring is fairly predictable, furthermore it is easily calibrated in a testing machine and can be used simply with a dial gauge as shown in figure 7.13. Structural uses occur in support skirts for vertical tanks, stiffening ribs on castings and in one instance an ocean racing yacht mast. Basically such arrangements have the merit of ease of manufacture but little else. They lack stiffness and, since the radial forces terminate in effect at a flat surface, they engender high bending stresses. The boat mast mentioned above broke quite early in its career and was replaced by a simpler, sounder form.

We shall consider only the bending action; direct tension or compression is usually negligible in comparison. Note that the I value of a curved member of I-beam or similar section is less than that of the equivalent straight member, see section 7.10, while the stresses are higher for a given moment.

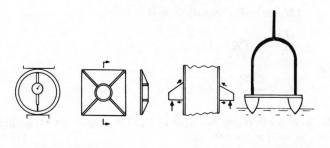

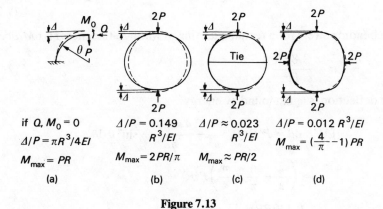

if Q, $M_0 = 0$ $\Delta/P = 0.149$ $\Delta/P \approx 0.023$ $\Delta/P = 0.012\ R^3/EI$

$\Delta/P = \pi R^3/4EI$ R^3/EI R^3/EI $M_{max} = (\dfrac{4}{\pi} - 1)\ PR$

$M_{max} = PR$ $M_{max} = 2PR/\pi$ $M_{max} \approx PR/2$

(a)   (b)   (c)   (d)

Figure 7.13

The analysis is based on one quarter of a ring. In the simplest case it is just a curved cantilever. In the second case the slope change $= 0$ over each quarter ring but lateral spread can take place freely. The third case assumes lateral spread to be prevented by rigid ties while the fourth, with equal loads, has diagonal axes of symmetry.

The simplest case has a load P only, M_0 and Q being zero (figure 7.13a).

The bending moment $= PR \sin \theta$. The deflection at P is found by energy: $P\Delta/2 = \int M^2 ds/2EI$ where $ds = R\,d\theta$ (see equation 7.5, etc.)

$$\int P^2 R^2 \sin^2 \theta\ R\,d\theta = P^2 R^3 \int \left[\frac{\theta}{2} - \tfrac{1}{4}\sin 2\theta\right]_0^{\pi/2} = \pi P^2 R^3/4$$

$$\Delta_1 = \frac{\pi PR^3}{4EI}$$

For the next case (b) we must assume an end moment M_0 so that the bending moment $M = M_0 - PR \sin \theta$. Since this time there is no slope change allowed, we use the moment area method $\int M\,ds = 0$ to give us the value of M_0; again $ds = R\,d\theta$

$$[M_0 R\theta + PR^2 \cos \theta]_0^{\pi/2} = 0$$

$$\frac{M_0 \pi}{2} - PR = 0$$

$$M_0 = \frac{2PR}{\pi}$$

Thus the general bending moment at θ is $PR\,(2/\pi - \sin \theta)$, giving a point of inflexion $(M = 0)$ at $\theta = 40°$

    The greatest bending moment is at $\theta = 0$, of magnitude $0.637\,PR$

$$(7.20a)$$

The bending stress for a rectangular ring of width b, thickness t is $6M/bt^2$

$$\sigma = \frac{3.8PR}{bt^2} \tag{7.20b}$$

The deflection is again found by energy

$$\int M^2 R\,\mathrm{d}\theta = P^2 R^3 \int_0^{\pi/2} \frac{4}{\pi^2} - \frac{4 \sin \theta}{\pi} + \sin^2 \theta\,\mathrm{d}\theta$$

$$= \left(\frac{2}{\pi} - \frac{4}{\pi} + \frac{\pi}{4} \right) P^2 R^3$$

$$\Delta_2 = \frac{0.149 PR^3}{EI} \tag{7.21}$$

The lateral spread is about $0.137PR^3/EI$ each side. If this is inhibited, the side-force developed is about $0.92P$ and the vertical deflection is only about $0.023PR^3/EI$.

    The symmetrical case (d) of four forces is solved thus: the forces on one quarter are P vertically and also P horizontally, with an end moment M_0. At any angle θ from the top

$$M = M_0 + PR(1 - \cos \theta) - PR \sin \theta$$

From the symmetry argument, the slope change $\phi = 0$ over limits 0 to $\pi/4$ or 0 to $\pi/2$, giving $M_0 = (4/\pi - 1)PR$. The highest stress $\sigma = 1.64PR/bt^2$ if the section is as above, b wide and t thick.

    The deflection is found by the energy integration $\int_0^{\pi/2} M^2 R\,\mathrm{d}\theta/2EI$ but we must bear in mind that this is the energy supplied by the work done at *both* ends of the quadrant

$$\text{The radial deflection} = 0.012PR^3/EI \tag{7.22}$$

bearing in mind that this is still in terms of the half-load, P.

    The deflections right across the ring under a full force which we can call W

will then have the same numerical values since $W = 2P$ but the deflection at both points is included; $2\Delta/W = 2\Delta/2P = \Delta/P$.
To summarise:

a whole ring under a radial force W deflects by $0.149WR^3/EI$;
when lateral expansion is completely prevented, it deflects by $0.023\ WR^3/EI$;
when loaded by equal forces W at right angles, it deflects by $0.012WR^3/EI$.

7.9 RECTANGULAR FRAMES

The two chief fields of use of rectangular frames are rolling mill end-frames and industrial buildings. The former have very simple loading, usually no constraint on side deflection but varying I values. Industrial buildings tend to use the same I value throughout a given frame but have various loadings due to snow, wind, cladding weight, crane rails, etc., and varying end constraints, fixed, pivoted or intermediate.

7.9.1 Frames with Free Sides

A typical rolling mill frame is shown in figure 7.14. The main loading is vertical but sometimes considerable horizontal forces are exerted on the slender side-bars due to entry or exit forces as the rolls start or cease to bite on the material, or due to uneven roll wear causing one roll to overtake the other. As the side-bars are usually longer and slimmer than the horizontals, the side-forces may be treated as loads on built-in beams, while the vertical loads are not very different from point loads on simply-supported beams.

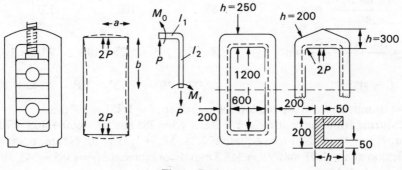

Figure 7.14

For the frame treatment we use one quarter. The angle change $\theta = \int M/EI$ ds = 0 over one quarter, by symmetry.

$M = M_0 - Px$ from 0 to a, then in the vertical, $M = M_0 - Pa$ over the length b (constant M)

$$\int_0^a \frac{M_0 - Px}{I_1} \, dx + \frac{(M_0 - Pa)b}{I_2} = 0 \quad \text{(moment area)} \quad (7.23)$$

$$\frac{M_0 a}{I_1} - \frac{Pa^2}{2I_1} + \frac{M_0 b}{I_2} - \frac{Pab}{I_2} = 0$$

$$M_0 = \frac{P\left(\dfrac{a^2}{2I_1} + \dfrac{ab}{I_2}\right)}{\left(\dfrac{a}{I_1} + \dfrac{b}{I_2}\right)} \quad (7.24)$$

The point of inflexion where $M = 0$ comes at $x = M_0/P$ and is generally close to the corner.

If the section varies with x then the recommended method is to create a function expressing $1/I$ in terms of x. This will be shown by an example.

Example 7.3

Find the stresses and extensions for the two frames shown in figure 7.14, made of material with $E = 2 \times 10^5$ N/mm² under a force of 600 kN (this is $2P$). All dimensions are in mm.

First we find the I values of the cross-sections as in section 2.2.2

h (mm)	200	250	300
Area = $100h + 5000$ mm²	2.5×10^4	3×10^4	3.5×10^4
1st moment = $\dfrac{100(h^2 + 50^2)}{2}$ mm³	2.125×10^6	3.25×10^6	4.625×10^6
$\bar{y} = \dfrac{\text{1st moment}}{\text{area}}$ mm	85	108	132
$I' = \dfrac{100 \, (h^3 + 50^3)}{3}$ mm⁴	2.708×10^8	5.25×10^8	9.04×10^8
$I = I' - A\bar{y}^2$	0.902×10^8	1.751×10^8	2.942×10^8

First frame, uniform top section 250 mm deep, $I = 1.751 \times 10^8$ mm⁴.
Measuring from the centroids, $a = 300 + 85 = 385$ mm, $b = 600 + 108 = 708$ mm; $P = 3 \times 10^5$ N. From equation 7.24, $M_0 = 1.03 \times 10^8$ N mm, point of inflexion at $x = 1.03 \times 10^8/P = 343.3$ mm. Foot moment $M_f = (385 - 343.3)/P$ $= 0.125 \times 10^8$ N mm, constant along whole side bar.

Stresses:
At $x = 0$

$$\frac{M_0 y}{I} = \frac{1.03(250 - 108)}{1.751} = 83.5 \text{ N/mm}^2$$

Mean shear stress $= \frac{P}{A} = 10 \text{ N/mm}^2$

In side bar

$$\frac{P}{A} + \frac{M_f y}{I} = 12 + \frac{0.125 \times 85}{0.902} = 23.8 \text{ N/mm}^2 \quad \text{tensile stress}$$

or

$$12 - \frac{0.125 \times (200 - 85)}{0.902} = 3.9 \text{ N/mm}^2 \quad \text{compressive stress}$$

Deflection: Tension in side bar over 600 mm length $= PL/AE = 0.036$ mm.
Shear in top bar $= PL/AG = \tau L/G = 12 \times 300/78\ 000 = 0.046$ mm.
Bending of top bar as cantilever up to point of inflexion (standard case)

$$PL^3/3EI = \frac{3 \times 343.3^3 \times 10^5}{3 \times 2 \times 10^5 \times 1.751 \times 10^8} = 0.116 \text{ mm}$$

From point of inflexion to corner, slope of horizontal $\times (385 - 343.3)$ plus cantilever of length $(385 - 343.3)$ as standard case

$$\theta = \frac{M_f \times 600}{EI}$$

by moment area

$$\theta = \left(\frac{0.125 \times 600}{2 \times 0.902}\right) \times 10^{-5}$$

$$\text{Deflection} = (385 - 343.3)\theta + 3 \times 10^5 \frac{(385 - 343.3)^3}{3EI} = 0.02 \text{ mm}$$

Total $= 0.22$ mm

For the variable frame we express $1/I$ as an integrable function, for example

$$10^{-8}/I = 0.34 + 0.77\left(\frac{x}{a}\right)^{1\frac{3}{4}} \quad \text{(highest value in centre)}$$

$\theta = 0$ if

$$\int_0^a (M_0 - Px)\left[0.34 + 0.77\left(\frac{x}{a}\right)^{1\frac{3}{4}}\right] dx + \frac{(M_0 - Pa)b}{I_2} = 0$$

$$0.34\,M_0 a - 0.17Pa^2 + \frac{0.77M_0 a}{2\tfrac{3}{4}} - \frac{0.77Pa^2}{3\tfrac{3}{4}} + \frac{M_0 b}{I_2} - \frac{Pab}{I_2} = 0$$

$$M_0 = \frac{Pa(0.375a + 1.109b)}{0.62a + 1.109b} = 0.908\,P$$

$$M_0 = 1.049 \times 10^8\,\text{N mm}$$

Point of inflexion is at $0.908 \times 385 = 350$ mm from origin.

Stresses:
At origin

$$\frac{My}{I} = \frac{1.049\,(300 - 132)}{2.942} = 60\,\text{N/mm}^2 \qquad \text{tensile stress}$$

Check at $x = a/2$

$$M = 1.049 - \frac{3 \times 385}{2 \times 10^3} = 0.4715\,\text{Nmm}$$

$$\frac{My}{I} = 0.471 \times \frac{142}{1.75}$$

In side-bar

$$M = (385 - 350) \times 3 \times 10^5 \qquad \text{stresses are 25 and } -1.3\,\text{N/mm}^2$$

Deflections would be calculated by energy $\int M^2/2EI\,\mathrm{d}s$; with the above expression for $1/I$ in terms of positive powers of x this is readily integrable.

If the top and bottom bars are different from each other, it is not valid to take a quarter-frame as basis; the integration has to extend from centre top to centre bottom.

7.9.2 Base-hinged Portal Frame

Next we consider a frame with uniform section, central load but with located ends. This corresponds to a rail-mounted crane gantry, for example. For analysis we fix the centre and consider the deflections of the half-frame, width B, height H. First we find the splay due to the up-force P alone. For small deflections P alone does not cause a bending moment in the upright (see figure 7.15). The end-slope is a standard cantilever case $PB^2/2EI$, so the splay at the foot $= PHB^2/2EI$. However, this splay is in fact prevented by the side-force Q. The deflection just found is equal and opposite to the deflection caused by Q alone; this consists of two parts: height H times slope change of

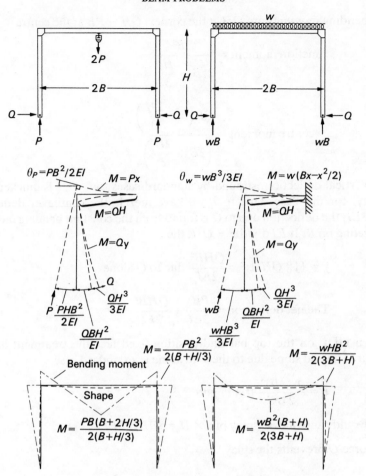

Figure 7.15

B, plus cantilever deflection of H due to Q. Moment due to Q alone $= QH$, $\theta = QHB/EI$ by moment area

$$\Delta = \frac{QH^2B}{EI} + \frac{QH^3}{3EI}$$

Making the deflections equal, $\Delta = PHB^2/2EI$, therefore

$$Q = \frac{PB^2}{2H\left(B + \dfrac{H}{3}\right)}$$

The bending moments are QH at the corner, $QH - PB$ at the centre

$$\text{Junction moment} = \frac{PB^2}{2\left(B + \dfrac{H}{3}\right)}$$

$$\text{centre moment} = \frac{PB\left(B + \dfrac{2H}{3}\right)}{\left(2B + \dfrac{2H}{3}\right)}$$

The vertical deflection is found by standard cases too. This is quicker than energy, for this case. Due to P we have merely the cantilever deflection $PB^3/3EI$; the deflection due to Q is found from the constant bending moment QH acting on B. If $EI\,\mathrm{d}^2y/\mathrm{d}x^2 = QH$, then

$$y = \int\!\!\int_0^B QH\,\mathrm{d}x^2 = \frac{QHB^2}{2EI} \text{ due to } Q \text{ alone}$$

$$\text{The net deflection} = \frac{PB^3}{3EI} - \frac{QHB^2}{2EI} \tag{7.25}$$

With a UDL on the top bar and position-fixed feet the treatment is very similar. The end slope due to the load alone is a standard case,

$$\theta = \frac{w(2B)^3}{24EI} \text{ from figure 7.3}$$

This would cause a base splay of $\theta \times H = wHB^3/3EI$.

The force Q prevents the splay

$$\text{deflection due to } Q \text{ alone} = \frac{QBH^2}{EI} + \frac{QH^3}{3EI} \qquad \text{as above}$$

therefore

$$Q = \frac{wB^3}{3BH + H^2}$$

The bending moments are QH and $QH - wB^2 + wB^2/2$.

$$\text{Junction moment} = \frac{wB^3}{3B + H}$$

$$\text{centre moment} = \frac{wB^2(B + H)}{6B + 2H} \qquad \text{(figure 7.15)}$$

7.9.3 Base-fixed Portal Frame

If the foot is place- and angle-fixed we need two equations, to find Q and the foot moment M_f. These arise from the horizontal deflection of the upright being zero and from the slope change from foot to centre being zero. Taking the origin at the foot,

$$EI \frac{d^2x}{dy^2} = M_f - Qy$$

$$EI \frac{dx}{dy} = M_f y - \frac{Qy^2}{2}$$

(notice dx/dy because we are integrating vertically)

$$EIx = \frac{M_f y^2}{2} - \frac{Qy^3}{6} = 0 \quad \text{when } y = H$$

thus $M_f = QH/3$; corner moment $= -2QH/3$.

The bending moment $M = QH/3 - Qy$ up to the corner, then $M = -2QH/3 + Px$ to the centre

$$\text{Moment area } \int M \, ds = \left[\frac{QHy}{3} - \frac{Qy^2}{2} \right]_0^H + \left[\frac{-2QHx}{3} + \frac{Px^2}{2} \right]_0^B = 0$$

$$\frac{QH^2}{6} + \frac{2QHB}{3} = \frac{PB^2}{2}$$

$$QH(H + 4B) = 3PB^2$$

$$\text{Foot moment} = \frac{QH}{3} = \frac{PB^2}{H + 4B} \tag{7.26}$$

$$\text{Corner moment} = \frac{-2PB^2}{H + 4B} \tag{7.27}$$

$$\text{Centre moment} = PB - \frac{2PB^2}{H + 4B} = \frac{PB(H + 2B)}{H + 4B} \quad \text{(figure 7.16)} \tag{7.28}$$

If the feet are place- and angle-fixed with a UDL, the relation of $M_f = QH/3$ still holds, hence the corner moment is still $- 2M_f$. The bending moment equaton is now $M = M_f - Qy$ up to the corner, then $M = -2M_f + wBx - wx^2/2$

$$\int M \, ds = M_f H - \frac{QH^2}{2} - 2M_f B + \frac{wB^3}{2} - \frac{wB^3}{6} = 0$$

Substituting $QH = 3M_f$ from above

$$M_f \left(\frac{H}{2} + 2B \right) = \frac{wB^3}{3}$$

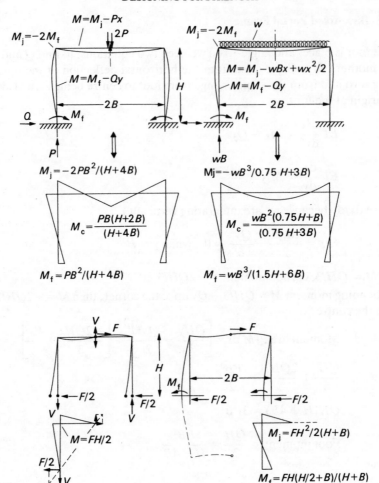

Figure 7.16

Thus we have

$$\text{foot moment } M_f = \frac{wB^3}{1.5H + 6B} \tag{7.29}$$

$$\text{corner moment} = -2M_f = \frac{-wB^3}{0.75H + 3B} \tag{7.30}$$

$$\text{centre moment} = \frac{wB^2}{2} - 2M_f = \frac{wB^2(0.75H + B)}{0.75H + 3B}$$

$$\text{(figure 7.16)} \tag{7.31}$$

The only further case developed here is the horizontal force. For small deflections only the line of action matters, not the point of application. Hence we apply the force F at the centre. By symmetry there is a point of inflexion here, the base reactions are equal to $F/2$ each. For pinned feet the case breaks down into four cantilevers; two with end force $F/2$, the other two with a notional vertical force $V = FH/B$.

The deflection at F is readily found by the strain energy of the four cantilevers made equal to $F\Delta/2$.

If the feet are fixed, then this half-frame is simply a pin-jointed case as above but lying on its side, the load being $-F/2$ and the pivoted foot represents the point of inflexion. The moments are given in figure 7.16.

In extending the treatment to sloping members the x, y coordinates may be maintained providing we remember that the fundamental equations are $\theta = \int M/EI \, ds$ and $\Delta = \int \theta \, ds$ where $ds = dx \sec \alpha$ or $dy \sec \alpha$ when the member is inclined at α to the relevant axis. In adding up deflections, we remember that Δ is normal to the member concerned and needs resolving into x and y components. A large number of ready-made solutions may be found in reference 35.

7.10 BENDING OF CURVED MEMBERS, INCLUDING FLANGE DISTORTION

This subject consists of two quite distinct aspects; solid and hollow bars. In solid bars there is stress concentration but not much change of cross-section. In hollow bars such as tubes (considered in section 9.9) and in I-beams and channels, the section changes are substantial, causing additional stresses and additional deflections.

First we consider rectangular solid bars. There are at least two analyses in the regular text-books giving different answers. Since they are not normally given in full it is not possible to see what assumptions they imply. A fundamental analysis with minimal assumptions is given in appendix A.9. The conclusions are shown in figure 7.17 in the form of an SCF, the ratio of highest bending stress to stress in a straight bar of the same section under the same moment. It roughly agrees with SCF $= 1 + \frac{3}{4} \ln t/r$. It agrees fully with one set of previously published values. If the direct force is small (load far from corner), it is near enough to add on the mean direct stress. If the load is near the corner, this alters the distribution. The equations in appendix A.9 deal with this quite simply. They also cope with tapered sections as in crane hooks.

The flexibility of a curved solid bar is rarely important, but just for interest it is given as a ratio, comparing the deflection with that of a straight bar of the same mean-line length m. The curved bar is stiffer because its neutral plane n

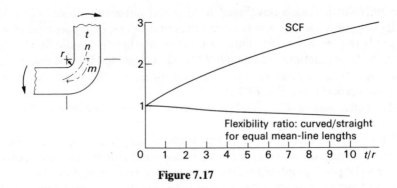

Figure 7.17

shifts towards the inner corner. This increases the second moment of area and shortens the effective length.

In I-beams the effects are mainly due to deflection of the flanges. The free edge of the tension flange tends to straighten out and the compression flange tends to curve more sharply. This greatly reduces the effectiveness of the flanges in resisting the bending moment and also produces stresses due to the transverse bending. The analysis given here is adapted from Anderson. [36] Figure 7.18 shows sections through a curved I-beam with and without stiffeners.

The radial shift is always in the direction which reduces the tension or compression. The change of radius is small enough to let us ignore the flexure stress lengthwise, so we consider the flange tension and compression, calling the stresses hoop stresses σ. The shift is resisted by the transverse stiffness; thus any flexure produces a transverse bending stress in the flanges, σ_t. Either of these can be the greater; in any case we need a reference stress to compare with. The logical reference stress is that produced by a given bending moment in the undistorted section, $\sigma_r = M/Z$.

The highest hoop stress is always greater than σ_r because the curved beam is always less than 100 per cent effective. The hoop stress comes to a peak at the points above the web or webs. The transverse stress σ_t may be higher or lower than σ_r, depending on conditions. It is liable to stress concentration where the web is joined to the flanges.

The analysis is too lengthy to be given in full. Basically, a hoop stress σ in a member of thickness t, curved to a radius r, produces a radial force per unit area of $\sigma t/r$. If the particular region is deflected radially by an amount y, then its stress is reduced below the maximum value we call σ_h

$$\sigma = \sigma_h - \frac{Ey}{r} \tag{7.32}$$

The radial pressure $p = \dfrac{\sigma t}{r} = \dfrac{\sigma_h t}{r} - \dfrac{Ety}{r^2}$ \hfill (7.33)

Looking at the protruding flange as a beam or cantilever, of unit width but part of a wide beam so that we should use $E' = E/(1 - \nu^2)$, while I for unit width $= t^3/12$, then by equation 7.7

$$\frac{W}{E'I} = \frac{P}{E'I} = \frac{d^4y}{dx^4} = \frac{12(1 - \nu^2)y}{t^2r^2} - \frac{12\sigma_h}{E'rt^2} \tag{7.34}$$

This is integrable, using sin, cos, sinh and cosh functions, with coefficients depending on the end conditions. A selection from Anderson's data is included in figure 7.18. It should be noted that Anderson amply verified his theory with strain measurements. Further data may be found in ESDU item 71004. [27]

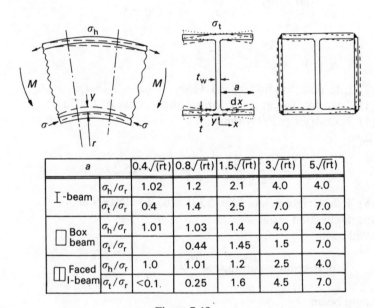

a		$0.4\sqrt{(rt)}$	$0.8\sqrt{(rt)}$	$1.5\sqrt{(rt)}$	$3\sqrt{(rt)}$	$5\sqrt{(rt)}$
I-beam	σ_h/σ_r	1.02	1.2	2.1	4.0	4.0
	σ_t/σ_r	0.4	1.4	2.5	7.0	7.0
Box beam	σ_h/σ_r	1.01	1.03	1.4	4.0	4.0
	σ_t/σ_r		0.44	1.45	1.5	7.0
Faced I-beam	σ_h/σ_r	1.0	1.01	1.2	2.5	4.0
	σ_t/σ_r	<0.1.	0.25	1.6	4.5	7.0

Figure 7.18

The deflection as compared with an undistorted beam is simply magnified in the ratio σ_h/σ_r.

It is important to note that the tangential stress is tensile on one face of the flange, compressive on the opposite. Thus there is sure to be one point where it is additive in the equivalent stress calculation. Equation 2.36 shows that the maximum shear stress is $\frac{1}{2}(\sigma_1 - \sigma_2)$ (neglecting direct shear), so the equivalent tensile stress in the table above is the *sum* of the given component stresses. The stresses are shown separately in case σ_h or σ_t is subject to stress concentration.

Example 7.4

A chassis member made from two pressings firmly fastened back-to-back is in effect an I-beam 120 mm deep (that is, high), 80 mm wide, flange thickness 5 mm, web thickness 10 mm. It is curved into a ½ metre radius at the mean line. Find the maximum permissible bending moment if the equivalent tensile stress is not to exceed 200 N/mm². Also find the bending moment to give this stress in a straight beam of the same cross-section and the ratio of the deflection of the curved beam to that of the straight beam under the same *moment*.

(1) Find $a/\sqrt{(rt)}$

$$a = \frac{(80 - 10)}{2} = 35 \text{ mm} \qquad rt = 500 \times 5$$

$$\frac{a}{\sqrt{(rt)}} = \frac{35}{\sqrt{2500}} = 0.7$$

(2) Interpolate to find $\sigma_h = 1.155\sigma_r$, $\qquad \sigma_t = 1.15\,\sigma_r$;
(3) Take equivalent $\sigma = 2 \times$ max. shear stress $= 2.305\sigma_r$;
(4) $2.305My/I = 200$ N/mm²,

therefore

$$M = \frac{200\,(120^3 \times 80 - 110^3 \times 70)}{12 \times 60 \times 2.305} = 65.2 \text{ kN m}$$

(5) In a straight beam, $M = 2.305 \times 65.2 = 150$ kN m
(6) Deflection ratio:

in curved beam

$$\text{deflection} = \text{constant} \times \frac{\sigma_h}{y}$$

in straight beam

$$\text{deflection} = \text{constant} \times \frac{\sigma_r}{y}$$

Therefore

$$\text{ratio} = 1.155 \text{ under equal moments}$$

It is advisable to check the web stresses for tension or buckling. The radial force $= \sigma_h \times$ flange area $\times \theta$, resisted by an area of $t_w r\theta$; hence the web stress must be about $\sigma_h \times 2at/rt_w$. This will only be important if the web is very thin and r is small.

Kinked beams are liable to lateral failure which combines bending and twisting. The analysis of straight beams in twist–bend buckling is not unduly difficult and is discussed in chapter 8; for kinked beams the problem is too advanced. The designer needs to be aware of it and provide lateral bracing. Cardboard models are recommended for detecting failure modes.

The chief design lesson is to keep the flanges narrow in curved beams and/or to support them with facings. These need not be continuous; separate strips or facings with lightening holes have the advantage of giving access for inspection and painting. An alternative form of reinforcement is radial gussets but these are unfavourable in fatigue since they put a stress-raiser right across the section. In castings radial gussets are acceptable and much easier to produce than facings; in welded structures gussets should only be used if fluctuations of stress are low; see section 8.5 for the stress-raising effect of welded attachments. The high values in figure 7.18 are not likely to occur in normal designs.

7.11 MAXWELL'S RECIPROCITY THEOREM

This theorem is found in most respectable textbooks. Owing to the rarity of its applications we can only devote limited space to it. The theorem states that in a linearly behaving structure the deflection at A due to a load at B is equal to the deflection at B due to the same load at A.

To test the theorem, consider a cantilever as in figure 7.2. For ease of notation let us make the length $= 1$ and $P/EI = 1$. Now we put a unit load at a distance x from the fixed end. The deflection at $x = Px^3/3EI = x^3/3$. To find the end deflection, we add $(1 - x)$ times the slope at x

$$\text{End deflection} = \frac{x^3}{3} + (1 - x)\frac{x^2}{2} = \frac{x^2}{2} - \frac{x^3}{6}$$

By Maxwell, placing the load at the end should produce a deflection at x of the same amount. Checking in figure 7.2 we find that the value there, arrived at by integration, agrees.

As a demonstration of the use of the theorem we shall verify the optimum spacing of figure 7.11c. For ease of writing we make the rolling load $P = 1$, $EI = 1$ and the span $= 2$. To find the bending moments we need the value of the reaction on the central support; calling this Q we know that the deflection due to Q alone which is $Q \times 2^3/48$ must equal the deflection at the centre due to P at some position z. Placing load P at the centre with Q absent, the deflection at z is given by figure 7.3 (noting that the span between supports $= 2$ for this purpose). By Maxwell this equals the deflection at Q due to load P at z, hence we can say

$$Q \times \frac{2^3}{48} = \frac{2^2 z}{16} - \frac{z^3}{12} \quad Q = 1\tfrac{1}{2}z - \tfrac{1}{2}z^3$$

This enables us to find the outer reactions by taking moments

$$2R_1 = (2 - z)P - Q \qquad 2R_2 = zP - Q$$

Substituting for Q and simplifying, remembering that $P = 1$

$$R_1 = 1 - \frac{5z}{4} + \frac{z^3}{4} \quad \text{and } R_2 = \frac{z^3}{4} - \frac{z}{4}$$

The bending moment at $P = R_1 \times z$, the bending moment at $Q = R_2 \times 1$.

Figure 7.19 shows the notation and the table below shows the results for various values of z.

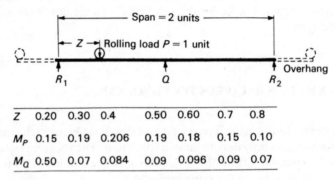

Z	0.20	0.30	0.4	0.50	0.60	0.7	0.8
M_P	0.15	0.19	0.206	0.19	0.18	0.15	0.10
M_Q	0.50	0.07	0.084	0.09	0.096	0.09	0.07

Figure 7.19

The highest value, 0.206, decides the size of beam required. We can therefore extend the beam beyond the supports until the moment with P at the end reaches the same value. The total beam length can therefore be $2 + 2 \times 0.206$ units. Applying this to figure 7.11c where the beam length L is fixed and we choose the best value for x, this is clearly $0.206L/2.406$ or $L/12$.

This result would take much longer to arrive at if we did not use Maxwell's theorem since this enabled us to use a standard beam case.

7.12 FLAT PLATES

Flat plates are a much neglected subject and tempt the designer to ignore the serious stresses which can be involved. We shall deal only with uniform loading on the whole surface and with loads terminating in a rigid central pad.

The values given apply only to small deflections, where umbrella action is negligible. At large deflections, radial and hoop stresses are set up in addition to bending action. Except in certain pre-stressed or pre-dished oil-can situations, the additional stresses stiffen the plate, reducing deflection and stress for a given load. These effects begin to show when the deflection exceeds half the thickness; when the deflection is about twice the thickness, the umbrella effects are as important as the bending effects, according to reference 5. The author does not vouch for this.

7.12.1 Simply Supported Circular Plate, Uniform Load p per unit area, Radius r (Diameter D) Thickness t

$$\sigma = \frac{3(3 + v)pr^2}{8t^2}$$

for $v = 0.28$ this gives

$$\frac{1.23pr^2}{t^2} = \frac{0.31\,pD^2}{t^2}$$

$$\Delta = \frac{3(1 - v)(5 + v)pr^4}{16Et^3}$$

if $v = 0.28$ this is

$$\frac{0.71pr^4}{Et^3} = \frac{0.044\,pD^4}{Et^3}$$

7.12.2 Clamped Circular Plate, otherwise as above

$$\sigma = \frac{0.75pr^2}{t^2} \quad \text{at edge,} \ = \frac{0.19\,pD^2}{t^2}$$

$$\Delta = \frac{3(1 - v^2)pr^4}{16Et^3}$$

if $v = 0.28$, this is

$$\frac{0.173pr^4}{Et^3} = \frac{0.011pD^4}{Et^3}$$

7.12.3 Simply Supported Square Plate, Side a, Uniform Load p, Thickness t

$$\sigma = \frac{0.22(1 + v)pa^2}{t^2}$$

if $v = 0.28$ this is

$$\frac{0.28pa^2}{t^2} \quad \text{if corners are held down}$$

$$\Delta = \frac{0.049(1 - \nu^2)pa^4}{Et^3}$$

if $\nu = 0.28$ this is

$$\frac{0.045pa^4}{Et^3} \quad \text{if corners are held down}$$

7.12.4 Clamped Square Plate, otherwise as 7.12.3

$$\sigma = \frac{0.31pa^2}{t^2} \quad \text{at mid-point of edge}$$

$$\Delta = \frac{0.014pa^4}{Et^3}$$

7.12.5 Rectangular Plates, Short Side a, Long Side b

Simply supported: based on square $a \times a$
for stress take

$$\text{square value} \times \frac{2.6}{(1 + 1.6a^3/b^3)}$$

for deflection take

$$\text{square value} \times \frac{3.2}{(1 + 2.2a^3/b^3)}$$

Clamped:
for stress take

$$\text{square value} \times \frac{1\frac{5}{8}}{(1 + \frac{5}{8}a^6/b^6)}$$

for deflection take

$$\text{square value} \times \frac{2}{(1 + a^5/b^5)}$$

7.12.6 Loads on Limited Area

This is a very important engineering case particularly in sheet metal work. The cases dealt with are simply supported and clamped round plates loaded through a rigid foot. Figure 7.20 shows the bending stresses and deflections

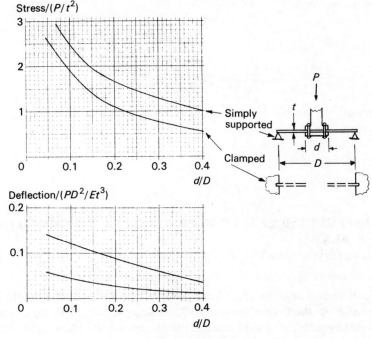

Figure 7.20

for a simply supported and a clamped plate of diameter D under a load P applied through rigid feet of varying diameters. The foot in each case is one that grips the plate firmly, not allowing it to bow within the fixing. With an applied couple M, the stress is roughly $4\ln(D/d) \times M/Dt^2$. Square plates may be treated as round plates of equal area. The data are from reference 5, calculated by the author using $\nu = 0.28$.

Example 7.5

A steel water tank side plate is 5 mm thick, simply supported by framing at 0.6 and 1 m spacing. If the mean pressure is 0.4 bar (0.04 N/mm²) find the stress and deflection. First work out values for 0.6 m square

$$\sigma = 0.28 \times 0.04 \left(\frac{600}{5}\right)^2 = 161 \text{ N/mm}^2$$

$$\Delta = \frac{0.045 \times 0.04 \times 600^4}{200\,000 \times 5^3} = 9.3 \text{ mm}$$

Adapt for rectangle

$$\sigma = \frac{161 \times 2.6}{1 + 1.6 \times 0.6^3} = 311 \text{ N/mm}^2$$

$$\Delta = \frac{9.3 \times 3.2}{1 + 2.2 \times 0.6^3} = 20 \text{ mm}$$

Comment: the simply supported condition is unduly pessimistic regarding deflection.

7.13 EFFECT OF HOLES AND BOSSES ON STRENGTH AND STIFFNESS OF BEAMS
(Fourth year work)

A hole in a beam near the neutral axis has relatively little effect if small except for its SCF in shear (see section 4.6). If large, it should be amenable to treatment by calculating the I value of the net section and allowing for an SCF as in figure 4.12, see figure 7.21a.

If the hole lies in the height direction its effect is more important. For stress we can assume the magnification to be that given by the SCF in tension from figure 4.12. If we wish to know the stiffness change, which may make a substantial difference to the behaviour of a frame, we shall make some simplifying assumptions.

Some region next to the hole, in line with the forces, is lowly stressed; by energy considerations it therefore contributes little to the stiffness or deflection. On the other hand material under high stress transverse to the forces is likely to contribute significantly since energy per unit volume varies as the square of the stress. The assumption made here takes an equivalent hole of parabolic outline, figure 7.21b. The law is $y = x^2/r = 2x^2/d$ which gives the correct radius of curvature at the narrowest section in order to simulate the high stresses correctly. The shaded area is taken as non-contributing. We wish to find an equivalent width b_e which has the same effect. The ratio applies equally to any layer parallel to the neutral axis; thus we compare the cases on the basis of unit height. Under any tensile force P the extension $= \int P/E(b - d + 2y) \, dx$ over the limits $\pm r\sqrt{2}$. It is easily shown that this is equivalent to a width $b - d/\sqrt{2}$ over a length $d\sqrt{2}$. The stress is simply the value of My/I using the equivalent width b_e, multiplied by the SCF of a bar with a hole, figure 4.12 or about $3 - 1\frac{1}{4}(d/b)^{0.6}$.

A more common form in press-frames, etc., is the enlarged boss with a hole, figure 7.21c. To find the equivalent we use section 7.8. A ring of radius r

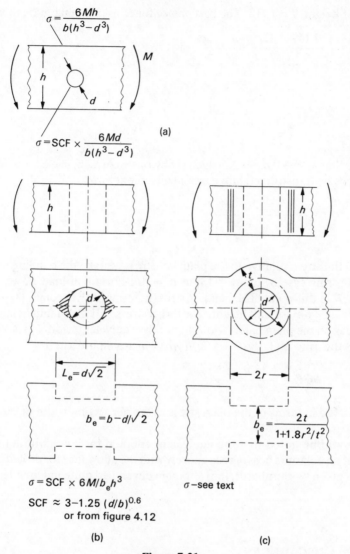

$$\sigma = \frac{6Mh}{b(h^3 - d^3)}$$

$$\sigma = \text{SCF} \times \frac{6Md}{b(h^3 - d^3)}$$

(a)

$L_e = d\sqrt{2}$

$b_e = b - d/\sqrt{2}$

$\sigma = \text{SCF} \times 6M/b_e h^3$

SCF $\approx 3 - 1.25 \, (d/b)^{0.6}$
or from figure 4.12

(b)

$2r$

$b_e = \dfrac{2t}{1 + 1.8r^2/t^2}$

σ —see text

(c)

Figure 7.21

under a force W extends $0.149Wr^3/EI$. A plain bar of width b_e, unit height, extends by WL_e/Eb_e where L_e, b_e are the equivalent length and width. The ring also has direct extension of the order of $W \times 2r/2tE$, per unit height. Thus

$$\frac{0.149r^3}{I} + \frac{r}{t} = \frac{L_e}{b_e}$$

For unit height, $I = t^3/12$. The best choice for L_e is presumably $2r$, therefore

$$\frac{0.149 \times 12r^3}{t^3} + \frac{r}{t} = \frac{2r}{b_e}$$

so

$$b_e = \frac{1}{\left(\dfrac{1}{2t} + \dfrac{0.9r^2}{t^3}\right)}$$

We see that as $r \to 0$ $b_e \to 2t$, which is obviously correct. The expression may therefore be transformed into a more convenient form

$$b_e = \frac{2t}{\left(1 + \dfrac{1.8r^2}{t^2}\right)}$$

To find the stress relative to the plain bar stress we take the bending moment from the ring case b, figure 7.13, at $\theta = 90°$. For a half-load P, $M = (1 - 0.637)PR$ at this point. The bending stress in terms of W becomes $(1 - 0.637) \times \frac{1}{2}W \times r \times 6/t^2$ for unit height. The direct stress $= W/2t$ for unit height while the stress on the plain bar of width b (not the equivalent width b_e) is W/b.

Thus the ratio is $(1.18r/t^2 + 1/2t)/(1/b)$ which simplifies to

$$SCF = \frac{1.18br}{t^2} + \frac{b}{2t}$$

If preferred in terms of the plain stress at the net section width of $2t$, SCF $= 1 + 2.4r/t$.

It should be noted that the equivalent adopted for a bar with a hole but without an enlarged boss agrees entirely with the allowance for drilled holes in flanges given by Bernhardt, [55] although developed quite independently.

7.14 PLASTIC HINGES

This subject has a growing application in structural design despite the major plastic modulus of the more efficient sections being only 10 to 20 per cent greater than the elastic modulus. Not all structures can be considered satisfactory when yielding since at yield the lateral stability tends to vanish, hence in structures designed on this basis we must provide adequate lateral support.

Cases of interest are (a) beams with a UDL and fixed ends and (b) portal frames. Suppose we have a beam over many supports so that the ends are effectively fixed by symmetry. Then we have end moments of $wL^2/12$, centre moments in each bay of $wL^2/24$. Elastic design would demand a beam of elastic modulus $I/y > wL^2/12\sigma_y$. If we adopt elasto-plastic design we would tolerate yielding at the supports but not quite allow yielding at the centre. The bending moment diagram would shift its base-line as in figure 7.22 until both moments are nearly equal. Thus we could reduce the beam section to a plastic

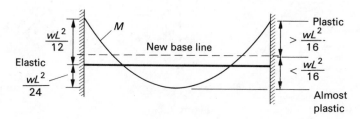

Figure 7.22

modulus of just over $wL^2/16\sigma_y$ without incurring gross yielding. If the beam is one of at least two and carries a floor slab, etc., to give lateral stability, then we could achieve a substantial saving of material provided the deflection is tolerable. At first sight the saving seems to be only about 25 per cent but when structures are so large that their main burden is self-weight, the saving becomes very much larger.

In portal frames, as we have seen, the moments at the various corners differ considerably so we can consider the frame to be safe when all corners except the last are yielding plastically, always provided that lateral constraints are enough to prevent out-of-plane collapse.

7.15 CENTRE OF FLEXURE, SHEAR CENTRE

Unsymmetrical sections have an odd property which if neglected may lead to unwanted torsional stresses and deflections. The most common case is a channel section, figure 7.23a. The effect of the shear reaction flowing from the web into the flange twists the free portion. To oppose this, the load should be applied at an outrigger of approximately $B/2$.

In angle sections the shear centre is near the corner. When using an angle section as a cantilever or beam, a curious effect is noted; the deflection is at an angle to the load. With an equal-legged angle it is almost at 45°. The reason seems to be the discrepancy of the moments of inertia. The highest is about four times the lowest so that the member behaves like a flat bar set up obliquely, as shown in figure 7.23b.

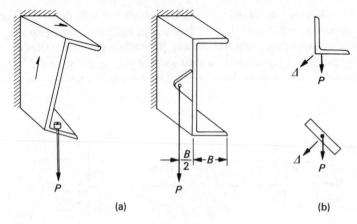

(a)   (b)

Figure 7.23

7.16 SPRINGS WITH BENDING ACTION

Here we discuss briefly multi-leaf springs, taper-leaf springs, clock and constant torque springs and torsion or mousetrap springs. There are of course many simple leaf springs used in low stress applications where cheapness and space saving are their great merits. These are simple cases of bars in bending as discussed in section 7.3.

7.16.1 Multi-leaf Springs

The most widely used version of this is the semi-elliptic vehicle spring, a simply supported beam usually with a central load. Occasionally we find a quarter-elliptic which is a cantilever. As a spring it is less economical since much material is inactive in the fixing but it has the advantage of not needing a compliant end-fitting. The length change between the ends is quite significant in leaf springs, requiring at least one end to be free lengthwise. This is achieved by a shackle, rubber mounting or slipper-guide.

The primitive theory of the multi-leaf spring is that the separate leaves act in unison so that in effect the spring is wider towards the point of higher bending moment; see figure 7.24a. This cannot happen exactly because the shorter leaves do not have a bending moment at the tips but a point load. The loading condition is shown in figure 7.24b.

Vehicle springs have further subtleties (c). The main leaf is normally used for location and is designed for relatively low stress in the normal working position; the vehicle is mainly carried by the other leaves. Sometimes the second leaf is formed as a reverse locator. The other leaves vary in curvature and thickness; sometimes the lowest leaves are fairly flat, coming into action only under heavy loads. The advantages of multi-leaf springs are that a broken leaf can be tolerated for a time, the material is not expensive, tolerances are not critical. The disadvantage is mainly the friction between leaves, leading to wear and rough action.

For calculation purposes we add the I value of each leaf at a given cross-section, then if the bending moment at that cross-section $= M, M/\Sigma I = E/R$. For each leaf, $\sigma = Ey(1/R - 1/R_0)$ where R_0 is the free radius of that particular leaf. To find the deflection of the whole spring by this means is tedious. As the design approximates to the constant stress condition (not throughout the material, but constant maximum stress at the surface of each leaf), the same expression as for taper-leaf springs, namely equation 7.35, should give a fair approximation. If the main leaf is at low stress in the laden position, it should be left out of account for the static deflection but included in stiffness calculations (P/Δ).

7.16.2 Taper-leaf Springs

These have just one leaf, or very few leaves, tapered so that the surface stress is approximately constant. This means that the thickness varies as the square root of the bending moment, except towards the ends where shear stress and longitudinal force demand a reasonable thickness.

If we can ignore the departure from theory towards the ends, the analysis is readily derived from energy by getting down to very basic terms. Under tensile stress σ the energy in unit volume $\sigma^2/2E$, strain times mean stress during the build-up. We use this in a rectangular bar, thickness $2h$, width b; therefore volume element $dV = bL\, dy$. $\sigma = \sigma_{max}y/h$. For a length L

$$U = bL\frac{\displaystyle\int_{-h}^{+h}\dfrac{\sigma_{max}^{2}\,y^{2}}{h^{2}}dy}{2E} = \sigma_{max}^{2}\frac{\dfrac{bL}{h^{2}}\left[\dfrac{y^{3}}{3}\right]_{-h}^{+h}}{2E}$$

$$U = \frac{\dfrac{2hbL\,\sigma_{max}^{2}}{3}}{2E} = \frac{\sigma_{max}^{2}V}{6E}$$

where V is the volume of metal $(V = \int_{-h}^{+h} bL\, dy)$.
But $U = P\Delta/2$, therefore

$$\Delta = \frac{\sigma_{max}^{2}V}{3\,PE} \tag{7.35}$$

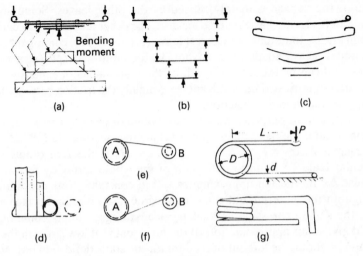

Figure 7.24

7.16.3 Clock Springs and Constant Torque Springs

Clock springs in the free state are of a self-generating form; in manufacture a strip of suitable yield point is wound onto a small spindle, causing the material to yield. It yields most at the centre, less in the outer coils. The free state is of little relevance since the starting condition is limited by the available space. When the spring is wound up, the stress depends on the yield point during forming, the effect of heat treatment if any, and the size of the centre. If possible, the working centre should be larger than the centre used in manufacture. Since the energy available depends critically on the yield point and the size of the housing, no general rule is given.

A variant is the constant torque spring. It is a proprietary article [37] which surprisingly has not found wide use. It can be made from any springy strip by rolling it over a three-roll system forming it to a constant radius of curvature. In use it is spooled on to a larger radius bobbin, or (in thin material) spooled on to a bobbin in reversed curvature; it works by its tendency to resume the curvature to which it was formed. The most familiar application is the constant force book clip (figure 7.24d). The energy change is a function of the change of radius and the length of strip involved in the change. The curled radius is almost constant, hence the energy change per unit unwound length or rewound length is also constant, hence force or torque is constant. Torque motor layouts are shown in figures 7.24e and f. The strip tries to get on to spool B when working; to wind it up we must get it on to spool A.

The energy of flexure of a length L is $M^2L/2E'I$. We transpose this by the bending equation $M/I = E'/R$; when the initial free radius is R_a, $M/I = E'/R - E'/R_a$ since force is constant, the force should be $\frac{1}{2}EI(1/R \pm 1/R_a)^2$, the minus case when spooling without reverse flexure, the plus case when reversing. [38]

The spring form shown in figure 7.24g takes the name torsion spring from the fact that the helix is loaded torsionally although the wire is in bending, by analogy with the compression spring where the helix is in compression while the wire is in torsion. Its advantages over a compression spring are lower inertia, no significant surge danger and compactness in the deflection direction. Its main drawback is the arc-shaped travel and inconvenient location in some installations.

The stress is a case of bending, maximum bending moment being $P(L + D/2)$. The stress is magnified at the inside of the curve. The values for rectangular bars from section 7.10, figure 7.17 should give an approximate value for the SCF due to curvature. The best loading direction is as shown since it makes the magnified stress compressive and the tension at the outer surface is diminished due to the direct compression. The force may be derived by energy

$$U = \frac{P\Delta}{2} = \frac{\int M^2 \, dl}{2EI} = P^2 \times L^2 \times \frac{\pi \, nD}{2EI} + \frac{2P^2L^3}{6EI}$$

$$\frac{\Delta}{P} = \frac{L^2}{EI}\left(\pi nD + \frac{2L}{3}\right) \tag{7.36}$$

Note that n in the version shown is $3\frac{1}{2}$, also if the lower arm is held down all along, we should substitute $L/3$ for $2L/3$.

PROBLEMS

First Year Standard

7.1 A springboard is mounted as in figure 7.4a, $a = 0.8$ m, $b = 3$ m. The board is 300 mm wide, 50 mm thick, $E' = 12\,000$ N/mm². Find the highest stress and deflection under a steady load of 75 kg at P.

Ans. 17.6 N/mm², 224 mm.

7.2 A trailer has two axles 5 m apart; it is 7 m long, overhanging 0.5 m at the front, 1.5 m at the rear. Find the highest bending moment under

    (a) a uniform load w per unit length all over;
    (b) a uniform load w per unit length spread over any part or parts;
    (c) a single load P;
    (d) two single loads Q.

Indicate the load position and bending moment position.

Ans. (a) 2.52*w*, 2.3 m behind front axle; (b) 3⅛*w* midway between axles, load between axles only; (c) 1.5 *P*, load at rear, moment at rear axle; (d) 3*Q*, both loads together placed as in (c); any load separation reduces moment.

7.3 A cantilever of length 2*a* has a download *P* at the centre. Find the upward force needed at the end to bring (a) the end, (b) the loading point *P*, up to the unloaded position.

Ans. 0.3125*P*, 0.4 *P*.

7.4 A beam of length 4*a*, with simple end supports has three equal loads *P* at *a*, 2*a* and 3*a*. Working from the centre, find the highest bending moment and deflection. Compare with those for a UDL of the same total spread out over the whole length.

Ans. 2*Pa*, 3⅙*Pa³*/*EI*, 1½*Pa*, 2½*Pa³*/*EI*.

7.5 A simply supported beam with a central load has reinforcement welded to the flanges so that its moment of inertia = 2*I* over the central half. Find the deflection as a ratio of the deflection of an unreinforced beam under the same loading. The notation *L* = 4*a* will be found convenient, using figure 7.25a.

Ans. 9/16.

7.6 A brass ring, *E* = 100 000 N/mm² is made of 5 mm diameter rod, the mean ring diameter being 50 mm. Find the radial force which will produce a stress of 60 N/mm², and the diametral extension of the ring.

Ans. 92½ N, 0.072 mm.

Advanced Questions

7.7 An I-beam of material with a tensile yield point of 240 N/mm² is 160 mm high, 80 mm wide externally, all thicknesses being 10 mm. It spans 2 m over simple supports. Find the load which will cause yielding to start with the load (a) central, (b) ¼ m from one end, (c) very near one end. Ignore web buckling, web crushing, fillet radii and stress concentrations. Use the web area method for web shear stress.

Ans. (a) 67.8 kN, surface yields, (b) 130.2 kN, web-flange junction yields, (c) 168 kN.

7.8 A cast iron sluice gate is required to withstand a pressure of 4 bar with a stress not over 50 N/mm². The design is ribbed as shown in figure 7.25a. Find the necessary thicknesses t_1 and t_2 (to be approximately 0.6 t_1) based on section 7.7, also the uniform thickness for the same stress. Estimate the reduction of uniform thickness possible if horizontal ribs are placed at 1 m and ½ m spacing.

Ans. 33, 20, 31.6 mm; reduction in ratio 0.916 to 1, 0.62 to 1.

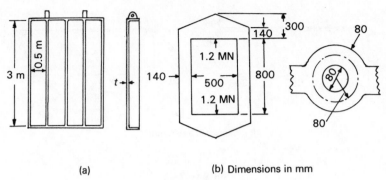

(a)   (b) Dimensions in mm

Figure 7.25

7.9 A frame as shown in figure 7.25b is 0.14 m thick. It is symmetrical horizontally and vertically.

(a) Confirm that the moment of inertia of the top and bottom bars is fairly represented by $I = 1/(3000 + 300\,000\, x^{2.6})$ m^4;

(b) Find the maximum stresses in the horizontal and vertical bars.

Ans. 68, 109 N/mm^2.

7.10 A long question. Repeat problem 7.10 with the top bar enlarged to form a collar for the loading screw, using the dimensions shown in figure 7.25c. Take an equivalent I as shown in section 7.13.

Ans. 194 N/mm^2 (SCF ≈ 2.94), ≈ 118 N/mm^2.

7.11 An engine connecting rod is 150 mm long between centres of the small and big ends. Its I/y value is 1100 mm^3. Treating it as a uniform rod results in an inertia loading acting transversely starting from 0 at the small end, to a value of 18 N/mm at the big end, increasing linearly. Using figure 7.6 find the reactions, then starting from the small end find the position of the maximum bending moment and the bending stress.

Ans. 900 N, 450 N, 86.6 mm from small end, M = 25 980 N mm, stress = 23.6 N/mm^2.

7.12 A form of vehicle suspension spring widely used is shown in figure 7.26a. The main leaf is 7 mm thick, the second 5 mm thick; both are 75 mm wide. Find the stiffness (force per mm deflection) at the ends (i) when the forces are both upward, (ii) when one force acts upwards, the other downwards (anti-roll action), assuming that both leaves remain in contact throughout. $E = 200\,000$ N/mm^2.

Ans. 16¼, 32½ N/mm.

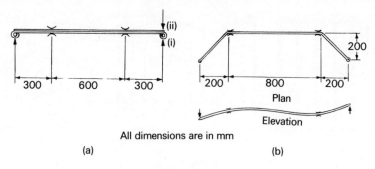

All dimensions are in mm

(a)   (b)

Figure 7.26

7.13   A steel anti-roll bar, 16 mm diameter, is shown in figure 7.26b. If the bar starts off in the horizontal plane, find approximately the force at one end to produce a vertical deflection of 90 mm. Assume the force to act vertically. Also find the highest Von Mises stress $(\sigma^2 + 3\tau^2)^{1/2}$. Consider torsion and bending but ignore stress concentrations. Treat the case as 45 mm deflections at each end, in opposite directions. $G = 78\ 000\ \text{N/mm}^2$, $E = 200\ 000\ \text{N/mm}^2$.

Ans. $\approx 700\ \text{N}$, $\approx 590\ \text{N/mm}^2$.

7.14   (Taken from a real fatigue failure, slightly simplified.) A sheet steel floor is 0.7 mm thick and is simply supported over a 400 mm diameter. What vertical force, distributed over a 100 mm diameter, will cause a bending stress of 250 N/mm²?

Ans. 83 N.

7.15   Figure 7.27 shows a roller carrying a conveyor belt exerting a load of 25 kN distributed uniformly over the central region of the roller as indicated by the arrows. The bearings are regarded as self-aligning, equivalent to simple supports. Find the bending moments and stresses at A, B and C, referring to chapter 4 for stress concentration data.

Ans. Moments 162.5, 750, 1750 N m; stress 50, 108, 42¼ N/mm².

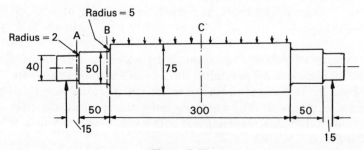

Figure 7.27

8 SOME ASPECTS OF STRUCTURAL STEELWORK DESIGN
(Third year standard)

To the general designer the task of designing large static structures seems simple; the functional requirement is merely that the object shall not fall down and shall not deflect excessively under load. The difficulties are these: firstly there is no chance of a prototype, the test bed is usually the general public; the second problem is that of economy. In a large structure most of the material has the job of supporting the other material. Even in railway trains and aircraft the dead-weight can greatly exceed the pay-load. In buildings and bridges any weight saved high up or at mid-span leads to extensive further savings.

Because of the lack of full-scale prototypes it is not safe to rely on the designer's individual interpretation of principles, loads and structural analysis. We work to well established rules concentrated into elaborate codes of practice such as BS 449, 153, 5400, the AISC handbook (American Institute of Steel Construction) and other associations' recommendations concerned with the safe design of road and rail bridges, high-rise buildings, etc. With care such rules can be used safely without much grasp of the underlying problems provided that the structures are free from unusual features. When innovation creeps in, so do disasters. This chapter aims to outline the chief problem areas physically.

Owing to the fragmented nature of this field which is due to the fact that we are pushing material economy towards the limit of safety in all directions, this chapter is somewhat difficult to follow. So indeed are the structural codes and textbooks.

If the safety factors mentioned in this chapter seem low, we need to bear in mind that they are based on loading values under severe conditions and on the yield stress. Hence a further safety margin exists in the material's ratio of UTS/σ_y.

8.1 ELASTIC LATERAL BUCKLING OF BEAMS (TWIST–BEND)

This form of buckling is three-dimensional, combining a down-load, lateral bending and torsion. These affect each other as shown in figure 8.1. When the

Figure 8.1

load exceeds a certain critical value the beam becomes metastable; a small lateral displacement causes the load line to be offset from the supports thus causing torsion. This causes twist which in turn means that the load has a lateral component acting upon the beam in the least rigid mode so that the lateral deflection increases further, etc.

The critical load depends on the elastic constants, the lateral moment and torsional stiffness of the beam section. For a beam of span L, reference 5 gives

$$P_{ce} = \frac{16\sqrt{(EI_y KG)}}{L^2} \tag{8.1}$$

E, G and I_y are defined in chapter 2. K is the torsional stiffness constant which is *not* the polar moment. For a rectangle, $K = (BT^3/3)(1 - 0.63T/B + 0.053T^5/B^5)$. For an I-beam, K is the sum of these values for the flanges and web taken separately.

This failure mode is rare in structural work just as elastic strut buckling is rare by itself. It could occur in narrow beams; for example floor joists in buildings may need to be stabilised by light diagonal braces; castellated roof beams are often of very slim profiles. This mode may contribute to the instability of long road trailers when negotiating roundabouts. When a narrow beam is used with high compressive stresses in one of the flanges, this buckling mode turns into a plastic failure mode, described in section 8.2.1.

A cantilever of length L should buckle when $P = \pi\sqrt{(EI_yKG)}/L^2$ (from Timoshenko). If the load point is above or below the neutral axis, the critical loads are decreased or increased, respectively (see reference 5 for extensive details).

8.2 BEAM FLANGE PROBLEMS

8.2.1 Lateral Bowing

This mode is closely related to twist–bend buckling in the same way as short column buckling relates to elastic buckling. It forms one of the limitations in beam design. To visualise the difference between twist–bend and bowing in the present sense, consider a cantilever which is quite stable in twist–bend mode but is loaded so that the compression flange is on the verge of yielding. An accidental side load could start the yielding at one point, then the deflection would induce a bending moment sideways, so the stress increases and yielding soon spreads right across.

In practical beam design we need to know the transverse radius of gyration and the effective end fixings. Figure 8.2 shows a convenient approach to the latter; if we take the *effective* length as shown, we use the pinned-end buckling

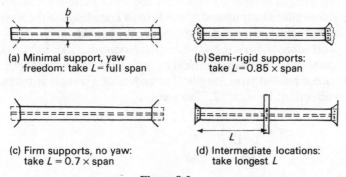

(a) Minimal support, yaw
    freedom: take $L=$ full span

(b) Semi-rigid supports:
    take $L=0.85 \times$ span

(c) Firm supports, no yaw:
    take $L = 0.7 \times$ span

(d) Intermediate locations:
    take longest L

Figure 8.2

formula, transposed into stress terms. This is done by dividing by the cross-section. If buckling force $= \pi^2 E I_y/L^2$ then buckling stress $= \pi^2 E I_y/AL^2$; the radius of gyration $k_y \equiv \sqrt{(I_y/A)}$, hence buckling stress $\sigma_{ce} = \pi^2 E k_y^2/L^2$. Note that L/k_y is often called the slenderness ratio. The value of k_y for a simple flat flange $= b/\sqrt{12}$; for tapered flanges it is found in beam tables (for built-up beams with edge stiffeners it is calculated as in chapter 2).

It is usually advisable to treat the beam compression flange as a short column, with interaction between elastic and plastic failure; failing stress $\sigma_f = 1/(1/\sigma_y + 1/\sigma_{ce})$. For example, in figure 8.2c, $\sigma_f = 1/[1/\sigma_y + (0.7L)^2/\pi^2 E(b^2/12)]$.

When working with the established design stresses, safe-load tables, etc., we can take it that provided there is no serious interaction with other failing modes the safety factor is about $1\frac{1}{2}$. The chief risk of slip-ups is with cantilevers; if these are free to wobble about sideways we must not use the beam values but either find safe loads for the cantilever condition or try the fundamental approach above. When using safe-load tables it is essential to read the instructions regarding lateral support assumed.

8.2.2 Wavy Buckling

This form of buckling was briefly mentioned in section 2.3.3. It may occur in thin plate girders and other thin structures such as aircraft fuselages, etc. In itself it is not a collapse mode though it can contribute to collapse since a wavy flange is only partly effective and may start overall yielding; it may also encourage bowing as in section 8.2.1 by reducing the effective k_y. The action is difficult to picture; figure 8.3 may help. Wavy buckling applies to struts as well as beams.

The characteristic wavelength of this mode depends on flange width. In low-yield rolled sections it does not normally occur because the stress needed is such that yielding starts first; it is a feature of high-strength steel sections with wide, thin flanges and plate girders built up from sheet and small sections. In flanges carried by two webs the wave pattern forms a quilt of square patches alternately up and down. From a knowledge of the wavelength we can estimate the stress needed to produce the buckles.

For a protruding width a the wavelength is somewhat like πa; this enables us to estimate the elastic critical stress by taking the half-wavelength $\pi a/2$ as the equivalent pinned strut length. The radius of gyration is taken in the thickness direction; $k = t_f/\sqrt{12}$. This gives

$$\sigma_{ce} = \frac{\pi^2 E \left(\dfrac{t_f}{\sqrt{12}}\right)^2}{\left(\dfrac{\pi a}{2}\right)^2} = \frac{E t_f^2}{3a^2}$$

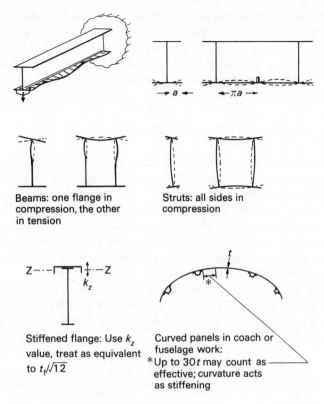

Beams: one flange in
compression, the other
in tension

Struts: all sides in
compression

Stiffened flange: Use k_z
value, treat as equivalent
to $t_f/\sqrt{12}$

Curved panels in coach or
fuselage work:
*Up to $30t$ may count as
effective; curvature acts
as stiffening

Figure 8.3

If we know the ratio E/σ_y then to ensure yielding before buckling we keep $a/t_f < \sqrt{(E/3\sigma_y)}$. Typically $\sigma_y/E = 1/1000$ for low-yield steels, 1/500 for relatively high-yield weldable structural steels and over 1/200 for some aluminium and titanium alloys. Substituting these values gives $a/t_f = \sqrt{(E/3\sigma_y)} = 18$, 13 and 8. The structural codes give permissible ratios of protruding width to thickness ranging from 12 to 16 for structural steels, with effective width between webs of about 40 thicknesses (this is basically π times the permitted protrusion). We can have wider flanges but these should be provided with stiffeners welded edge-on, to prevent the formation of long-wave buckles. As it is difficult to give a theoretical requirement for the size of stiffeners, the minimum proportions are given in the rules.

Wavy buckling can occur in beams or in columns, regardless of whether bowing is also possible or is prevented by sufficient side support.

Very few codes or textbooks consider interaction between wavy and lateral (bowing or twist–bend) buckling directly, relying on the separate require-ments. If B/t_f is inside the stated limit for the grade of steel and the compressive stress is inside the permissible value for a strut of the appropriate slenderness ratio L/k_y then this is considered sufficient in practice. We should bear in mind that the codes also put overall limits on L/H, L/B, H/B, B/t_f regardless of stress. When all these are close to the limit it may be wise to consider possible interaction (see section 8.4).

8.2.3 Shear Lag, Loss of Effective Width and Area

Finally we look at an important but much neglected aspect, important in aircraft wings, wide beams generally, notably box girders. We recall that the web of a beam transmits the shear force to the flange. The shear cannot stop suddenly at the junction. It disperses through the flange, falling to zero at the edge (at the centre in a box girder). The shear gradient is usually shown as linear, with uniform flange tension. This is near enough true in long narrow beams; in wide beams the dispersion is gradual, the tension and compression falls off asymptotically towards zero.

To assess this situation we consider a rectangular element of flange in a wide box girder; we choose a region where there is appreciable shear. The element is under tension σ at x, $\sigma - \delta\sigma$ at $x + dx$. The sides are in shear τ at z, $\tau - \delta\tau$ at $z + dz$. (See figure 8.4).

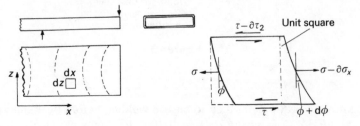

Figure 8.4

We consider first the balance of forces, second the connection between changes of stress and changes of shear angle. From force balance, $\delta\sigma_x\,dz = \delta\tau_z\,dx$, hence

$$\frac{\delta\sigma}{dx} = \frac{\delta\tau}{dz} \tag{8.2}$$

Under stress, the edge nearest the web is at higher stress and extends more. Length difference $= \delta\sigma_z\,dx/E$. Over the length dx, the shear angle increases by $d\phi$, therefore

$$d\phi = \frac{\text{length change}}{\text{width}} = \frac{dx\,\delta\sigma_z}{E\,dz}$$

By the shear law, $\phi = \tau/G$, hence $d\phi = d\tau/G$. Eliminating $d\phi$

$$\partial\tau/\partial x = (G/E)\,\partial\sigma/\partial z \tag{8.3}$$

The solution will depend on how the shear force varies with x. The tensile stress is likely to diminish exponentially across the section. Since the bending moment determines the *average* tensile stress, the maldistribution will give rise to excess tension at the corners, with possible start of yielding. The same maldistribution occurs on the compression side and may cause unexpected high stresses which could start a plastic collapse, plates and stiffeners both yielding together.

Some visual evidence of this type of action is seen in figure 8.5, from the little-known work of Kloth. [39] The white lines are contours from brittle lacquer data and presumably indicate the greater principal stress distribution. The graphs are strain-gauge results. In the lower view, the flat part is in compression; however as the stress is below 50 N/mm² there should be little buckling effect since this is well below the critical level. In this structure the strain-gauge data seem to be mainly related to local flexure. The main point is

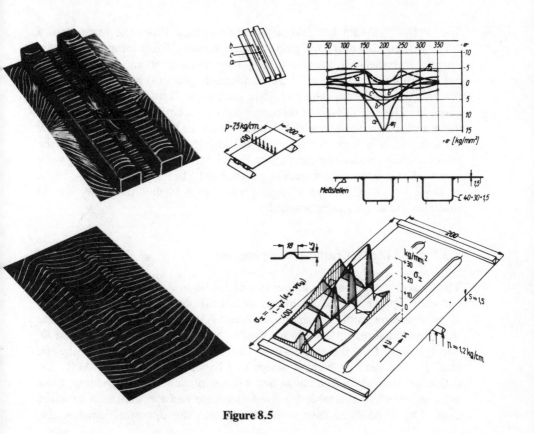

Figure 8.5

that plane sections do not remain plane. Incidentally, in the fluted panel there seems to be some yielding near mid-span; without this, the effect would be even more pronounced.

The simplest way of treating the loss of strength is the effective width coefficient. A set of coefficients appropriate to wide bridge girders is given in reference 40, clause 6.3.2. This appears to refer only to the tension flange; obviously the situation is similar to compression and indeed is more serious since a low effective width means higher stresses than might be supposed, which in turn could mean onset of buckling. If we get the shear lag wrong in tension, there will be considerable readjustment due to yielding and work-hardening so that the margin between yield and UTS acts in our favour; in buckling this aspect may not be available. Effective width coefficients can be as low as 0.18.

8.3 BEAM WEB DESIGN PROBLEMS

The web of a beam has several buckling modes. First, the force from a concentrated load or from a support reaction puts a direct compressive stress into the web. Second, the web is part of the bending situation and therefore has a longitudinal stress $\sigma = My/I$, increasing from 0 at the neutral axis towards the maximum value as it joins the flange. Third, the shear stress can cause buckling in a diagonal direction, forming a corrugated pattern. There are many occasions when all these modes reach the highest stress values in the same place. It is important to grasp the physical aspects of these since the codes are not necessarily clear about their interaction. Fourth, sloping webs are weaker in compressive buckling than may be thought at first sight. Fifth, at large lateral deflection a sloping web has a further instability due to nearness to the instantaneous centre.

8.3.1 Local Loads, Crushing, Crippling, etc.

The local effect has various names as indicated in the sub-title. Under the load shown in figure 8.6a, the conditions are not as severe as in a column since the stress decreases from the maximum value at A to zero at B. The codes assume that the shaded region is a column, the dispersion angle θ varying between 30° and 45°. So we treat the web as a column of width b, thickness t_w under the end load P. The next problem is which end conditions to assume. AISC distinguishes between a free flange and a flange prevented from twisting; these are equivalent to the table-leg fixed–free case and the unbraced structure case. The fixed–fixed case would imply that the opposite flange is also

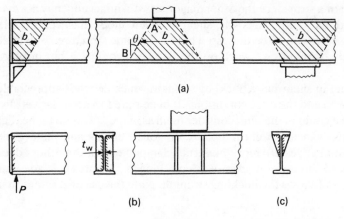

Figure 8.6

position-fixed. This would often be the case at the ends of a beam but rarely in the centre, so for safety no advantage is taken of this possibility. The empirical rules and recommendations take care of both the very local stress just under the loaded flange and the buckling risk of the whole web as the load spreads out. If the load is beyond the capacity of the web alone, stiffeners must be provided; again the rules seem to be straightforward so there is no need to devote further space to this problem.

Some stiffener types are shown in figure 8.6b. The form shown in c is an appropriate way of stiffening under fatigue loads predominantly in one direction. It stiffens the flange against wavy buckling and the web against crippling under a point load while leaving the tension side unaffected. There is a lack of relevant data; reference 41 reports briefly on some work in which the stiffener was carried through into the tension half. This turns it into a low-class welded detail, category E in figure 8.12 or class F_2 in figure 8.12, of considerable stress concentration. A more sensible version which stops at the neutral axis does not seem to have been investigated enough. The tensile half may not need stabilising.

The stiffener not only increases the area thus reducing the compressive stress; it forms a much more stable form of strut. The radius of gyration is that of a section comprising the stiffener or group of stiffeners plus about twenty thicknesses' worth of web.

8.3.2 Shear Buckling

A piece of beam web between stiffeners has to transmit the shear force; without the web, the flanges and stiffeners would fold up very easily. If the web is too thin, it will buckle diagonally. This is not an absolute failure like the buckling of a strut since the other diagonal is still intact and may not even have exceeded its yield. Thus the shear buckled mode has appreciable post-buckling strength, to be discussed below. In some structures web buckling is accepted. Normally we work on the basis of avoiding it but not by a wide margin.

A panel in shear has a buckling strength which depends appreciably on its aspect ratio and the edge conditions. If panels are fixed to relatively twistable frames we could in the limit consider them as hinged. This might be realistic in ship hulls, van and rail-car bodies. In rolled beams and plate girders the longitudinal edges are more like truly clamped edges, the other edges being somewhere inbetween clamped and hinged. Figure 8.7 adapted from reference 42 shows the buckling strength. Note the use of E rather than G.

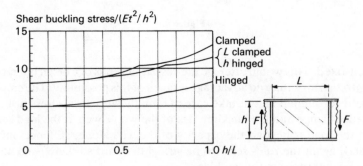

Figure 8.7

If we are working with unusually thin webs and wish to avoid relying on the tension-field in the other diagonal direction, we may stiffen the web diagonally, producing a mixture between a plate girder and a framed structure; the web is still useful in stabilising the compression members. Unfortunately the author is not aware of any design guides for such structures.

In BS 153 the longer side of a panel must not exceed $270t$, the shorter $180t$. Inside these limits, the shear stress must not exceed certain tabulated values.

Example 8.1

Let $h = 140t$, L between stiffeners $210t$. From the clamped–hinged graph

$$\tau_{ce} = 9.7E \left(\frac{t}{h}\right)^2 = 9.7 \times \frac{200\,000}{140^2} = 99 \text{ N/mm}^2$$

The stress from BS 153 = 121 N/mm² (table 11, steels of σ_y 400 to 450 N/mm²). [If we use the fully clamped graph and E', we get 10.45 × 200 000/140²(1 − 0.28²) = 116 N/mm².] So if we design to the standard we should expect some buckling, even with perfect clamping.

Estimate of Post-buckling Strength up to Yield Point, Onset of Gross Distortion

Consider one panel, height h, length L, thickness t; figure 8.8. Buckling strength $= \tau_{ce}$

$$\text{Strain energy in buckled mode} = \frac{\tau_{ce}^2}{2G} \times \text{volume}$$

when whole panel is buckled

$$\text{Strain energy in tensile mode} = \frac{\sigma_y^2}{2E} \times \text{volume}$$

when whole panel is yielding

Deflection at $P = \Delta$

$$\text{Energy} = \frac{P\Delta}{2} = \left(\frac{\tau_{ce}^2}{G} + \frac{\sigma_y^2}{E}\right)\frac{hLt}{2}$$

Obtain deflection by triangle as in section 5.1, regarding tensile mode as a diagonal member. From figure 8.8, $\Delta = h\cosec^2\theta\,\sigma_y/E$

$$P = \left(\frac{\tau_{ce}^2}{G} + \frac{\sigma_y^2}{E}\right)\frac{Lt\sin^2\theta\,E}{\sigma_y} \quad \text{(from the energy statement above)}$$

Transposing this into the form of a limiting strength $\tau_{\lim} = P/ht$

$$\tau_{\lim} \approx \sigma_y\left(1 + \frac{E}{G}\frac{T_{ce}^2}{\sigma_y^2}\right)\frac{L\sin^2\theta}{h}$$

But $L/h = \cot\theta$

$$\tau_{\lim} \approx \sigma_y\sin\theta\cos\theta\left(1 + \frac{E}{G}\frac{\tau_{ce}^2}{\sigma_y^2}\right) \tag{8.4}$$

It must, however, be admitted that Basler's test data in *Welding Research Council Bulletin 64* (1960) fit much better to the simple expression $\tau_{\lim} \geq \frac{1}{2}\sigma_y \sqrt{(\sin\theta)}$, even for unusually thin webs (derived from tests on life-sized steel plate girders 12 m long and 1.27 m deep).

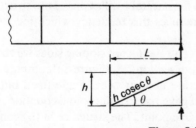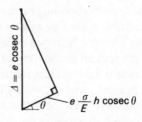

Figure 8.8

8.3.3 Longitudinal Buckling

Since a beam web is under longitudinal stress My/I, there is a possibility of buckling into vertical folds. It is a non-critical mode since a beam is often designed to take the moment entirely by the flanges, therefore a loss of support from the web lengthwise should not matter. Perhaps the main argument against permitting such buckling is that it might interact with flange wavy-buckling.

It is not easy to see a simple theory for this mode. Half the web is in tension; the part in the highest compression is near the flange and well supported by it. If we argue, say, that the lowest third of the web is relevant, then the effective strut length may be deduced from the concept of waviness in square-patch quilt pattern.

Let the panel size be $\frac{1}{3}h$. If we wish to ensure yield before buckling, and taking low-yield material of $\sigma_y = E/1000$, then the elastic critical stress relates to the stress in the relevant piece of *web* when the beam is yielding at the *flange*. In other words we shall compare σ_{ce} with about $\frac{3}{4}$ of σ_y.

$\sigma_{ce} = \pi^2 E k^2/(h/3)^2$; $k = t/\sqrt{12}$ as usual, and we shall set $\sigma_{ce} = \frac{3}{4}$ of $E/1000$ as discussed above

$$\frac{\pi^2 E t^2}{12\left(\dfrac{h}{3}\right)^2} = \frac{\frac{3}{4}E}{1000} \qquad \left(\frac{h}{t_{\lim}}\right) = 99$$

We note that BS 153 demands longitudinal stiffeners at $0.2h$ above the flange if h/t exceeds 155 to 200, depending on the yield point of the steel, and a further stiffener at the neutral axis if h/t exceeds 190 to 250, again depending on material. This implies a buckled-panel size of $h/6$. AISC rules tolerate this form of buckling, since fabricated girders are made strong enough to stand the bending moment on flange area alone.

8.3.4 Sloping Web Instability

Many modern bridges use trapezoidal box girders. These have a specific advantage regarding wind-induced vibration. Upright sides are prone to alternate eddy shedding in phase with the natural oscillations of the structure; a streamlined leading edge greatly reduces this tendency which led to the famous Tacoma Narrows collapse.

To present an assessment of the structural effect of sloping sides we review and expand the buckling treatment from section 2.3. Figure 8.9 compares the basic pinned strut (a) with the structure able to sway, angle-fixed but not position-fixed. We recall that the strut gave rise to a sinusoidal equation, with maximum deflection at the quarter-wave point. The structure in (b) consists of two quarter-wavelengths assembled with a point of inflexion at the node.

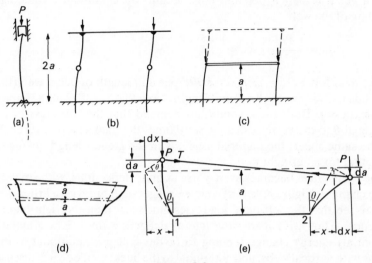

Figure 8.9

We imagine a bar with pin-joints across these nodes (c), to simplify the argument. We consider only small deflections, otherwise we would have to take note of load transfer from one strut to the other. The buckling strength of this arrangement is given by $P_{ce} = \pi^2 EI/(2a)^2$. We compare this with the sloping side version (d), again with a tie-bar across the nodes. At (e) we show the structure without the upper part. The flange plates are treated as rigid. If they are significantly flexible in some cases, this could be allowed for by assuming the webs to be slightly longer. Since the top half obeys similar arguments, the bar simply replaces the top half. Taking moments about the base points 1 and 2

$$M_1 = T(a + \mathrm{d}a) - P(x - \mathrm{d}x) \qquad M_2 = P(x + \mathrm{d}x) - T(a - \mathrm{d}a)$$

The total disturbing moment $= M_1 + M_2 = 2P\,\mathrm{d}x + 2T\,\mathrm{d}a$

From balance of forces, $T = P \tan \theta$ from geometry, $\mathrm{d}a = \tan \theta\,\mathrm{d}x$. Therefore

$$M_1 + M_2 = 2P(1 + \tan^2 \theta)\,\mathrm{d}x = 2P \sec^2 \theta\,\mathrm{d}x$$

In the upright structure, $\tan \theta = 0$, $M_1 + M_2 = 2P\,\mathrm{d}x$.

The strut force $= P \sec \theta$ (by resolving vertically). The strut length $= a \sec \theta$ from the point of inflexion to the fixing. The elastic crippling force along the web slope $= \pi^2 EI/4(a \sec \theta)^2$. Let us call the critical download P_{ce}, where $P_{ce} \sec \theta = \pi^2 EI/(2a \sec \theta)^2$. For equal stability when compared with $\theta = 0$, New I = old $I \times \sec^3 \theta$, or to express this more elegantly, in a sloping web box beam with a vertical load P per unit edge length, the minimum I value against buckling of the web

$$I_\theta = \frac{Ph^2}{\pi^2 E'} \sec^3 \theta \tag{8.5}$$

The critical force $P_{ce} = \pi^2 E' I \cos^3 \theta / h^2$ per unit length of beam web (that is, each side). $P_y = t\sigma_y \cos \theta$ per unit length. Note that $I = t^3/12$ for a simple web of thickness t. Both these conditions demand that the thickness of a web sloping at θ to the vertical shall be $\sec \theta$ times the thickness of a vertical web for the same shear; the material usage, with the greater length, being $\sec^2 \theta$ times that for an upright web.

The same effect becomes even worse at finite deflections, with appreciable angle changes. Figure 8.9e is of course exaggerated, but with a little goodwill we can see that the left-hand load rises much less than the right-hand load descends. Thus with increasing lateral deflection there is a gravitational (potential) energy change tending to favour collapse additional to that in beams with vertical webs, and additional to the buckling mode just discussed.

8.4 INTERACTION PROBLEMS

With so many different modes of failure it is difficult to obtain clear guidance on the severity of interaction. In normal applications the various overall maxima given in the codes and revised from time to time restrict L/H, L/B, B/H, B/t, H/t, etc., to a sufficient extent. This allows us to limit the amount of checking. Of course we shall check for yielding under combined stress by formulae given in the codes or by equations 2.37 to 2.40, or by Mohr's circle, (appendix A.11), remembering maximum in-plane and out-of-plane shear. The worries arise if we have high stresses from several modes coming together in the same spot.

Under the down-load we have the following possibilities for a given load and span:

(1) Twist–bend elastic buckling governed by I_y and torsional K but not directly by stress;

(2) Lateral flange bowing governed by L/B and stress;

(3) Wavy flange buckling governed by B/t_f and stress;

(4) Compressive yielding longitudinally governed by I_x and H, that is, by stress;

(5) Compressive yielding vertically governed by load width and t_w;

(6) Compressive buckling vertically governed by load width, t_w and H or h;

(7) Shear buckling of web, governed by t_w, h and L or stiffener spacing, and shear force;

(8) Shear yielding of web, governed by t_w and H, that is, web cross-section, and shear force;

(9) Longitudinal buckling of web governed by t_w and H and bending stress;

(10) Reduced permissible stress due to cyclic loading.

It has even been argued that if a perceptible deflection occurs under mode 5, the flange under longitudinal compression bows downwards and puts a further compressive force into the web.

There is no need to despair immediately. In most cases mode 1 will be prevented by the rest of the structure; likewise mode 2, except in gantry cranes. If so, we have no incentive to keep B large, we can use narrower, thicker flanges and thus avoid 3. 5, 6 and 7 may call for local stiffeners unless there is an objection from the fatigue viewpoint. (The partial stiffener shown in figure 8.9c is not yet widely established; if carried too far into the tension side it may cause trouble, if kept too short its effectiveness is not fully known.) 4 and 8 are the basic minimal requirements anyway; if the rest are reasonably provided for, the extra material needed in the flanges and webs to cover interaction, however calculated, will not be great.

For each of these modes we can find a critical stress once we have settled the provisional dimensions, k values, etc.; let us call these σ_{c1}, σ_{c2}, etc., (some will just be σ_y). Find the stresses under the working load, for each mode. These we call σ_1, σ_2, etc. Dividing each of these by the relevant critical stress we obtain ratios r_1, r_2, etc.

Now we require two things – a suitable safety factor and a suitable combination rule. Some rules say let the working stress $\sigma_w = \sigma_y \sqrt{(r_1^2 + r_2^2 + r_3^2 + \ldots)}$/safety factor. Another way is to treat modes 1 to 4 as parallel paths to disaster, obtaining an equivalent $r_{1234} = 1/(1/r_1 + 1/r_2 + 1/r_3 + 1/r_4)$.

A similar parallel seems advisable for modes 7 and 8 also including mode 4 in the web. Reference 40 gives data for this in graph form.

Modes 5 and 6 are automatically treated in this way because the rules for deducing safe working stresses for web compression use standard data for short columns. If these are not available, refer to section 2.3.3 and use a safety factor of at least 1.6, preferably 3.

If we satisfy these groups separately with a reasonable safety factor, this should be sufficient, for extra safety we can combine the three grouped ratios by the square root method and apply the safety factor at the end.

When using the code rules, it may be wise to make extra checks when, for example,

(1) L/B and B/t_f are both near the limit and general stress levels are high;

(2) shear-lag maldistribution, high compressive stress and high shear coincide;

(3) sloping webs come under high local forces or side force, particularly during erection;

(4) flange edges are damaged during handling and B/t ratios are near the limit;

(5) rolled sections are visibly distorted perhaps due to excess residual stress.

One of the problems with start of buckling is residual stresses. Rolled sections and welded fabrications are both liable to have the centre in tension, the outer edges in compression since the inner parts are the last to cool and contract. The localised compressive stress could start premature yielding. Gaylord and Gaylord [43] quote various test results showing that effect.

8.5 JOINTS AND JUNCTIONS

Inevitably joints and junctions are weaker than continuous members. Gross loss of continuity gives weakening in static loading as well as in fatigue; local yielding affords surprising amounts of relief in static cases, much less so in fatigue. To illustrate the kind of discontinuity that causes static weakening and how to alleviate it, figure 8.10 shows a few extracts from Haigh. [44] Joints 1 and 2 show an I-section cantilever welded to an I-beam upright with fillet welds and full penetration butt weld. The performance is similar, showing that the weld detail has not made much difference to the yield strength. The weakness is due to lack of continuity at the anchorage of flange A. The provision of continuity plates of various designs overcomes this weakness, as shown by the remaining joints.

The effect is greater at corner joints, showing an almost three-fold gain by using plates to support the unbalanced force at the flange tips, from an experiment carried out by the author on 3 in × 1½ in I-section steel. With the corner plates in use, the strength became substantially equal to that of the basic section. Adequate continuity or corner plates can almost completely eliminate the weakness. The design of corner plates is based simply on the triangle of forces – the flange of the I-beam has to transmit a certain force, obtained from the bending stress and flange area. The two flange forces FF give rise to a reaction R. The total cross-section of the corner plates, plus a contribution from the web, must be made ample for this force, allowing for any weakness of the welds at the same time. This should apply to any joint angle, not just 90°.

Under fatigue loading the reader may be surprised to find that in bridge work the stress range is used without the mean stress. One reason for this is the large scatter-band of the results; a more rational reason is that welds would generally not be stress-relieved so that a peak residual stress of σ_y is

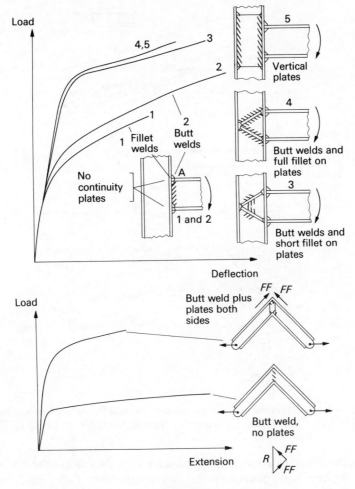

Figure 8.10

likely to exist in a weld. After loading, some further yielding is likely to occur locally; then as loading fluctuates we are working in the top corner zone of the Smith diagram (figure 3.14). Very conveniently all the usual structural steels can be treated together. The stronger steels may have a higher endurance limit on small smooth specimens but this is accompanied by greater notch sensitivity associated with finer grain size; moreover in the heat-affected zone (HAZ) next to a weld, the higher yield points tend to drop down to a common level. Figure 8.11 is a composite from reference 45. It shows the degree of scatter obtained from nominally similar assemblies, and a set of permitted

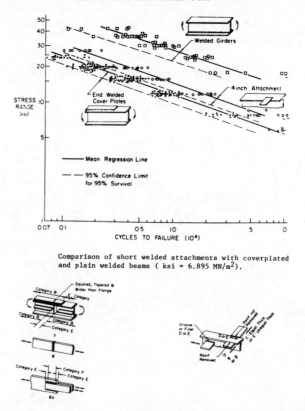

Comparison of short welded attachments with coverplated and plain welded beams (ksi = 6.895 MN/m^2).

Figure 8.11   Reproduced. by permission of Plenum Publishing Corporation, from J. B. Burke (ed.), *Application of Fracture Mechanics to Design* (1975)

stresses presented as a σ–N graph. These are U.S. data, hence the stresses are in ksi, thousands of pounds (force) per square inch. Category A is plain members without stress concentration. The ratios of the various stresses at a given life expectation are in effect the reciprocals of the stress concentration factors.

If the details seem hard to follow, some help may be got from the similar details of classes specified in part 10 of BS 5400, [46] see below.

In view of the wide scatter of results the distinction between classes seems somewhat artificial; as more evidence collects, some of the details become re-classified. It is wise to consult the most recent issue of the standard we are working to.

Under steady loading, stress levels in full-penetration butt welds can be up to the general permitted stresses, subject to good crack-free welding. The stresses in fillet welds are calculated in the traditional manner assuming that the lines of tension run through the fillet at 45° to the main force direction, hence the stress for this purpose is force/(length × throat size).

If the load is not applied symmetrically, it is converted into a force acting through the centroid of the weld group plus a couple. We find the highest force per unit length of weld as best we may and then take the stress as the force per unit length divided by throat thickness. There is some uncertainty about how the forces are shared out among various parts of a weld group, depending on relative elasticity of the welds and members. Hence this is not an accurate science and it is not usual to distinguish between shear and tensile stress in a fillet weld.

The traditional design stress for fillet welds in mild structural steel was 0.3 σ_y; BS 153 permits a stress of 0.5 σ_y for steady loads. The relevant parts of BS 5400 for steady loads are not available at time of going to press.

The approach in BS 5400 is shown with the help of figure 8.12. In addition to describing the method of calculation, the frequency of different loads to be

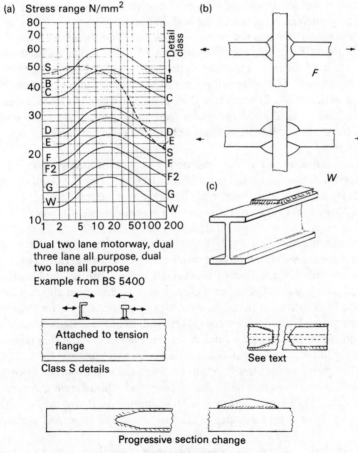

(a) Stress range N/mm²

Detail class

Dual two lane motorway, dual three lane all purpose, dual two lane all purpose
Example from BS 5400

Attached to tension flange

Class S details

(b)

F

W

(c)

See text

Progressive section change

Figure 8.12

expected, etc., it shows prepared stress values for various classes of road bridge under the usual loadings, having already taken into account the σ–N relation and Miner's law. One such set of stress values is shown in figure 8.12(a). The ratio of strength of the various details may be deduced directly from this graph. Class A is a machined test piece; plain continuous sections rank as class B. If such a section has holes in it of corner radius greater than the thickness *and* if we apply the recognised SCF then we may still use class B stress levels provided the edges are as rolled. If the member is a plate with cut edges, down it goes to class C. The code covers nearly all the known forms of weakness; only a small selection can be discussed here.

It must be noted that not only the shape is defined but also the fabrication arrangements. For instance a full penetration butt weld may be of class C if made to the most expensive standards, with run-on and run-off plates to avoid stop-starts, dressed flush, with minimal misalignment, welded from both sides, whereas under other conditions it may only class as D or E, or even down to F if welded from one side or if it joins I-beams where the residual stresses are much more serious than with flat plate.

The most instructive is the weakest class, W. One member of this class is shown in figure 8.12. The force has to pass through an intermediate plate in the direction of weakest plate strength, the unwelded centre represents a pre-existent crack. If it were full penetration, it would move up into class F. This type of detail is only permitted if we can ensure that the plate is free from lamellar defects. In BS 153 the incomplete penetration butt weld was only allowed for shear stress loading (part 4 clause 4.11); in tension *or compression* it was barred.

Another weak detail is the end of a beam reinforcing plate (c). This combines the following: a fairly sudden enlargement at the point of highest stress in bending, a fillet weld with dubious root penetration, and a heat-affected zone, all right across the section. This may not matter when the beam is lowly stressed, for example, if designed for stiffness. If it does matter, the designer can alleviate matters by tapering the end. Then the force is spread out over a longer weld run and the other effects are spread out so that at any one section there is a region of unaffected metal. The research data seem mostly negative; a point at 30° or a point which is blunt over the middle half, where the force comes in from the web, is not significantly better than a plain end. Tapered ends need to be strongly tapered, as narrow as practicable at the end. It has been shown that a further substantial stress reduction is obtained by thinning the material at the tip, before or after welding.

The strange curve for class S details may be seen in terms of an uncertainty factor; this class refers mainly to load-bearing attachments on the maximum tension face. This construction is not normally used for major loadings anyway, so there is no difficulty in making the detail large and the stresses small.

Since the fatigue stress is assessed by range regardless of mean stress, the question arises whether a detail can fail in fluctuating compression. It is difficult to visualise a compressive crack; the code leaves this problem to the designer's discretion. Under repeated severe compression it is possible for shear cracks to occur.

8.6 JOINTS IN TUBULAR STRUCTURES

This subject has come to the fore in connection with steel tube offshore platforms. Many of these use tubular steel legs with bracings made from smaller pipes welded onto them without local reinforcement. Local reinforcement as used in pressure vessels could be inappropriate because we are concerned with push–pull forces so that it may well be better not to interfere with the inherent flexibility of a plain joint. The main stresses are flexural stresses at the change of direction, with small contributions from direct stress in the branch and bending stress in the main pipe. Some recent data come from references 47 and 48.

For some years it was customary to present test results and design recommendations for joints in terms of the punching shear stress in the main pipe. Data quoted in Gurney, [41] pp. 190, 191, show this basis to be somewhat dubious, since on a punching shear basis it would seem that steel fails at 4 to 12 N/mm^2, varying strongly with size for geometrically similar joints.[†]

Fundamentally the flexure stress depends on the local bending moment where the two pipe walls meet. If d/D is very small, the clamped flat plate case of figure 7.20 would apply. At larger d/D values the deciding features would be where the effective point of inflexion comes and how the force distributes itself, mostly at A rather than B, figure 8.13. This seems to be a job for the computer and indeed some programmes have been developed. It is difficult to judge which effects have been ignored or oversimplified in such cases, so we look at the basics.

If we knew the position of the point of inflexion, we could get the bending moment and stress at the junction. The recognised flexibility parameter for curved shells is $\sqrt{(RT)}$ so we expect the distance to the point of inflexion to be more or less proportional to this. The section modulus is proportional to peripheral width $\times$ thickness^2. Then there should be a progressive effect due to increasing d/D ratio. Reber [47] shows an expression deduced from computer values

$$\sigma_{max} = 0.62PD^{0.7}T^{-1.6}d^{-1.1}$$

If we look at the implications we note that the emphasis is much as expected,

† Converting these test data to the Reber analysis only brings them up to 25 to 75 N/mm^2. One cannot help suspecting that a totally different factor is involved, for example, flat-plate conditions as per section 7.12.

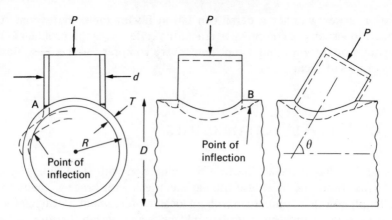

Figure 8.13

since we could break it down into items of $\sqrt{(RT)}$, $1/T^2d$ and $(D^2/Td)^{0.1}$ as the adjustment factor for branch size. It is clear that this will give stress in any consistent set of units since the dimensions are force/length². Applying this to the specimen used in reference 48, a well-documented set of stress measurements, gives a stress of twice the measured value. Yet in Gurney [41] a comparison shows that in some cases the Reber stress agrees well with measurements.

For a sloping joint at an angle θ, we would expect two effects; one is that the normal force, which presumably is the most important component, is reduced in the ratio sin θ. The other effect is that the effective periphery over which the flexure is spread is increased. Reber takes the overall effect as a multiplier of $\sin^{1.5} \theta$, showing that this fits in well with static tests to failure.

8.7 BOLTED OR RIVETED CONNECTIONS

8.7.1 Tensile Connections

The type of connection shown in figure 8.14 is subject to maximum stress rules that vary somewhat by code and type of duty; it behoves us to check whether we are working to building or crane or bridge rules. Basically a certain tensile stress must not be exceeded on the thread root area; this varies according to the grade of bolt material and the size. Smaller bolts, 18 mm and below, are restricted to lower stress values. There are two good reasons for this – danger

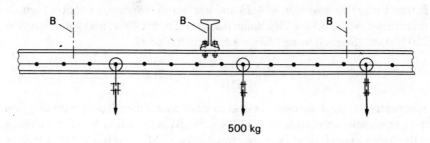

500 kg

Figure 8.14

of overtightening and loss by corrosion. Rivets should be used mainly in shear.

Designwise, there are two important aspects. If we are working with tapered flanges we must use girder washers to avoid bending the bolts. Moreover we should use the right type; there are at least two types, for I-beams and for channels, with different angles. The other point is not the bolts but the holes. Do we need to make an allowance for the reduced cross-section? Often we can place the holes where the bending moments are small, so the problem may not arise. The answer is to look at our data. If we are using safe load tables from AISC, these contain an allowance for 15 per cent loss of flange cross-section. The property tables in appendix A.14 show a gross and a net I value for I-beams. The net value allows for appropriate holes, as declared at the foot of the first set of tables. [48]

Example 8.2

Let figure 8.14 represent a conveyor carried on beams B at intervals; the load on each beam is 150 kg due to the rail and chain plus 500 kg intermittent loading due to the cargo. The beam spans 3 m (normal to the paper), is simply supported at the ends and loaded centrally. It is an 8 in × 4 in joist to BS 4, elastic modulus $I_x/y = 225.8$ cm³, flange width 102 mm, flange thickness (5° taper) = 10.4 mm average, web thickness 5.8 mm, $k_y = \sqrt{(I_y/A)} = 2.25$ cm (note BS tables use mm for dimensions, cm for properties). The beam and bolts are of steel of UTS 400 N/mm², σ_y 249 N/mm², endurance limit ± 180 N/mm².

The objective is to specify four suitable bolts and check the beam for safety.

Loading: we assume that not every conveyor is laden, so the load on the beam and joint may cycle between 150 and 650 kg an infinite number of times.

(1) Bolt selection
Load = 650 kg maximum, 150 kg minimum. (Amplitude ± 250 × 9.81 N, mean 400 × 9.81 N.)

Either find ratio max/min = 4.33 and use Smith diagram, or ratio of amplitude/mean = $^5/_8$ and use Goodman diagram. From either, maximum stress = 250 N/mm². Suggest using SCF = 4 as in reference 49.

$$\text{Net bolt area} = \frac{4 \times 650 \times 9.81}{250} = 100 \text{ mm}^2 \quad \text{each bolt 25 mm}^2$$

Alternatively, if we assume preload to yield point, then from Smith diagram the permissible amplitude = ± 130 N/mm². By section 6.6, bolt force change is 0.3 times load change, bolt area required with SCF 4 is 0.3 × 250 × 9.81 × 4/130 = 24 mm² for four bolts. We shall choose M12 bolts for general ruggedness and credibility.

Comment: traditional tensile design stress for small bolts in crane work = 36 N/mm²; we have 22 N/mm² nominal stress.

(2) Check for elastic twist–bend failure.
Failing load by equation 8.1 = 188 kN. This is an underestimate since the load is applied below centre which adds to stability and the bolting-up inhibits twist to some extent;

(3) Check for flange bowing
This is likely to be significant since in structural tables the highest span for this profile is 4¼ m. σ_{ce} from section 8.2.1 = $\pi^2 E k_y^2 / L^2$ = 111 N/mm². Note use of k_y from data, not from B; this is because joist sections are strongly tapered, so that $k_y < B/\sqrt{12}$. This is somewhat overcautious since it ignores stiffening due to the junction;

(4) Check for flange wavy-buckling
This is unlikely to be serious in a relatively narrow joist section. σ_{ce} from section 8.2.2 = $E t_f^2 / 3a^2$. As a worst assumption take $a = B = 102$ mm, σ_{ce} = 693 N/mm² (probable value 1000 to 1200 N/mm²).

(5) Combined failure check in compression flange
Twist–bend can be ignored since strength is vastly greater than other modes. Take the possible failing stresses as the fatigue and elastic failing stresses

$$\sigma_f = \frac{1}{\left(\dfrac{1}{240} + \dfrac{1}{111} + \dfrac{1}{693}\right)} = 68.4 \text{ N/mm}^2$$

$$\text{Actual } \sigma \text{ at maximum load} = My/I = \frac{\left(\dfrac{650 \times 9.81 \times 3000}{4}\right)}{225\ 800}$$

$$= 21.2 \text{ N/mm}^2$$

(6) Check for stress in tension flange
Nominal tensile stress from above = 21.2 N/mm². This is increased due to loss of cross-section at holes and SCF. From figure 4.12, SCF = 2.4. Cross-section

reduced from $t \times 102$ to $t \times 76$, having drilled two 13 mm holes for bolts, including clearance. Hence

$$\sigma = \frac{21.2 \times 2.4 \times 102}{76} = 68.3 \text{ N/mm}^2$$

(7) Check web tension
Assume width of joint = ¾ of the flange width = 76 mm. Web area = 76×5.8 = 441 mm²

$$\sigma = \frac{650 \times 9.81}{441} = 14.5 \text{ N/mm}^2$$

(8) Check web shear
Shear force to each side = $650 \times 9.81/2$ N; area = $180 \times 5.8 = 1044$ mm²; $\tau =$ 3 N/mm²;

(9) Answer
Use four M12 bolts, drill 13 mm (clearance) holes, all perfectly safe, factor *at least 3* in all modes examined.

The static safety factor in the bolts at yield = 11.2 ignoring SCF. Customary safety factor (based on UTS) = 6 for bolts, about 2.5 on structural members.

8.7.2 Shear Connections

The most ancient way of joining sheets of metal is riveting or bolting. For wide work it has been largely superseded by welding; elsewhere it is a valuable method of permanent or demountable construction, notably in heavy civil engineering and light aircraft work. In general mechanical work, riveting has the advantage of combining cheapness, high security and repairability. The mode of action is shear strength of the rivet or bolt, in the original sense of resistance to cutting by a pair of shears, aided by friction grip. In the extreme case of high-strength friction-grip bolting, the whole load is borne by friction; shear is a second line of defence.

The penalties are the extra material needed to provide the overlap and the weakening of the parent plate by holes. Empirical design is done by observing limits of shear stress and bearing stress and providing specified amounts of edge and rear distance; we shall discuss the action on which the rules are based.

Most joints use load sharing by two or more bolts or rivets; often we use symmetry, with the bolts or rivets in double shear, each shank being used to resist load on two planes simultaneously.

For simplicity we start with a single rivet, attaching a loaded bar to a structure of ample strength and ignoring any frictional support. A numerical example is easier to follow than general equations and makes for realism in comparing the various design possibilities, figure 8.15.

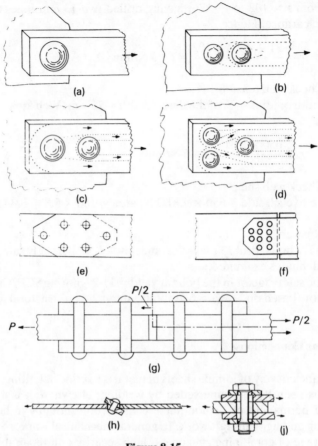

Figure 8.15

Example 8.3

Investigate joint designs to fail by fracture or gross yielding at 10 kN (limit design). Bar material $\sigma_y = 333 \text{ N/mm}^2$, rivet material $\tau = 200 \text{ N/mm}^2$.

(1) Find rivet area: 10 000/200 = 50 mm². Diameter of single rivet = 8 mm. (Figure 8.15a.)

(2) Find bar area: 10 000/333 = 30 mm², at smallest cross-section $(b - d) \times t$

Consider various thicknesses

t	$b - d = 30/t$	b	full area
mm	mm	mm	mm²
1	30	38	38
2	15	23	46
3	10	18	54

The reader can continue the table. We see that the most economical design is the thinnest. Why not use ½ mm, 68 mm wide? Because it would pull off over the rivet head.

(3) Decide a suitable thickness: a useful criterion is to keep the nominal contact stress between rivet and bar to about σ_y (reference 48). We can exceed σ_y locally because the contact is like in a hardness test in which the crushing stress is found to be over twice σ_y. With this rule, $d \times t \approx 30$ mm^2, $t = 3\frac{3}{4}$ mm if available, $b = 16$ mm. In practice we would make $b > 2d$ to avoid side weakness due to stress concentration, bulging due to rivet pressure during hammering-up, etc. We also provide a generous rear distance to avoid the kind of failure shown in figure 8.16. Edge and rear distances (hole centre to plate edge) in large structural steel plate work are 2 to 2½d; this allows for irregular placing, forcing into place, etc. In relatively accurate machine work, 1¼d at the sides, 1½d at the rear is usually enough.

(4) Now consider two rivets, (figure 8.15b). For 50 mm^2 area, $d = 5.656$, say 6 mm. Thickness by the contact stress criterion: $t = 30/2d = 2.5$ mm, or 3 mm (as available). Width is obtained again by $(b - d) \times t = 30$ mm^2; $b = 18$ mm if $t = 2.5$, or $b = 16$ mm if $t = 3$. Note the saving in cross-section.

(5) Next comes a more subtle point. Obviously we must maintain the minimum cross-section nearest to the load; beyond the first rivet the load is shared by both parts, so at the second rivet we only need enough cross-section for the load taken by the second rivet; the first part has already gone to ground. In other words we have excess material at the second rivet (this point is slightly more complex when joining equal members). To make fuller use of our material, we could use a smaller rivet near the load (rarely seen) or three rivets, as in figure 8.15c and d. The dotted lines give an impression of the way the load is shared. This principle is widely used in practice (e, f).

(6) For our load condition, the rivet area is still 50 mm^2; three equal rivets of 4.6 mm will provide this. Making it up to 5 mm standard size, the next step is to find the thickness. $3d \times t = 30$; $t = 2$ mm minimum, $b = 20$ mm. This type is more applicable when using small diameter high-strength rivets.

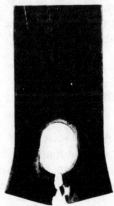

Figure 8.16

(7) Finally we must consider how to deal with double shear. This is best seen in figure 8.15g by considering just one symmetrical quarter; this of course has only a half-load, 5 kN in our example; from there on the calculation is as before, according to the number of rivets, but with the 5 kN load. The cover-plate is usually thicker than $t/2$.

(8) Recommended design stresses for mild steel rivets and bolts:
Shear 80 N/mm²; if holes are well aligned and rivets are hammered up under favourable conditions, 100 N/mm². [48]

Bearing stress calculated as nominal contact stress P/ndt, that is, force over number of rivets times diameter times plate thickness); reference 48 allows 250 N/mm²; this leads to very thin plates; the author prefers 180 to 200 N/mm².

We note that the shear stresses are well below the material strength, thus a safety factor has been included whereas in the bearing stresses there is much less margin. This on the face of it is acceptable since crushing is not a collapse mode; it is therefore important not to use this rule blindly. If the plate thickness is thin compared with rivet diameter, experience shows that we invite other failures if an overload occurs, for example, plate bending, stretching and pulling off over rivet head (figure 8.15h).

When using bolts, we need to consider whether the whole bolt shank is available to resist the shear or only the thread root area. This will vary from case to case.

(9) Friction grip bolts, figure 8.15j.

High-strength bolts with heavy-gauge washers, tightened to yield point, are used successfully in civil engineering construction. On as-rolled unpainted surfaces, with any loose rust or scale carefully removed, they form a very strong connection, overcoming the weakening effect of the hole. Grip forces of 150 N per mm² of bolt shank area can be relied on in such cases at each contact face (excluding the washers). It is essential to use only the approved high-tensile waisted-shank bolts; [48] these have enough resilience to maintain the grip despite any slight settlement of the surfaces. In general machine design, friction grip is depended upon to provide rigidity but no definite quantitative data can be given since smoothness and lubrication vary considerably.

PROBLEMS

8.1 A steel joist 254 × 114 mm as tabulated in appendix A.14 spans 7.5 m. What central load is liable to cause twist–bend failure? The ends are held upright but not otherwise restrained. $E = 210\,000$ N/mm², $G = 80\,000$ N/mm². Ignore flange taper, use mean thickness from table.

Ans. 26.6 kN.

8.2 In the same joist, what is the stress and deflection under a load of 15 kN assuming a drilled flange? Use net I value (for stress) but ignore SCF.

Ans. 84 N/mm², 12 mm.

8.3 Still in the same joist, if there is no lateral support, at what load would flange bowing failure become a danger? Take $\sigma_y = 240$ N/mm², use I_y/A.

Ans. 4 kN.

8.4 A universal beam 254 × 146 mm × 31 kg/m (see tables in appendix A.14) spans 3.6 m. The ends are supported so that *laterally* the effective free length = 0.85 × 3.6 m. $E = 210\ 000$ N/mm², $\sigma_y = 420$ N/mm². What central load may be carried with a safety factor of 1.5, based on

(a) separate criteria, that is, weakest single mode;
(b) reciprocal interaction;
(c) root and square interaction?

Ans. 103, 44.5, 72.5 kN.

9 PRESSURE VESSELS
(Third year standard)

9.1 INTRODUCTION

Pressure vessels should be easier to design than the structures described in chapter 8; their basic function is not to blow up or leak; a certain amount of yield is tolerated, this is described as shakedown. Moreover the shapes and details are highly standardised and well proven by previous service data and near full-size specimen tests. Weight as such is not critical. The pressure to economise, though laudable within limits, is aggravated by the custom of estimating cost for tendering purposes on a weight basis.

The so-called unfired pressure vessels as distinct from storage tanks and pipes are sometimes spherical, more usually cylindrical with domed end covers. The pressures are usually large compared with the static gravity pressures. The stresses in thin and thick cylinders and in thin spheres were covered in sections 6.3 and 6.4.3. In pressure vessels the problems are due to joints, end cover shapes, branch openings, localised forces at supports, etc. In some ways a pressure vessel has a lighter duty than other structures. For example, the working load is highly predictable since overpressure is prevented by safety devices, notably bursting discs which are domes made to a predictable bursting pressure. These are safer than spring-loaded safety valves since they are unlikely to get sealed by corrosion and are not liable to deliberate maladjustment. It is not uncommon however for these discs to be installed in reverse, or wrongly supported; this may make them effectively *too* strong. Also, every vessel is tested to over-stress before going into service. This, if carried out well, not only assures initial soundness; it also imposes residual stresses generally in a favourable direction, work-hardens the material locally and if tested up and down several times, monitored by suitable strain gauges, etc., it can show up incipient cracks. It is sometimes argued that the cracks are caused by the test; this story is more popular with the maker than with the customer or the insurer.

Pressure vessels are not normally designed from first principles; this would allow too much room for opinion and argument. Instead, sets of rules have been developed which started as very simple formulae, generally over-cautious within previous experience but which could become unsafe when extended into unusual regions. Over the years these rules have grown into codes of practice covering numerous eventualities and variations. They are now so elaborate that the risk of errors is considerable and the cost of designing is high.

Computer packages have become available for some aspects of pressure vessel detail design. This tends to eliminate clerical errors (though not punching-in errors) but has the effect of concealing the physical basis even more than the code layout does already.

To save space and confusion we only discuss the ASME boiler and pressure vessel design code section VIII divisions 1 and 2 [49] and the current British Standard BS 5500. [50] For brevity we call these A1, A2 and B. Note that these codes are growing plants, revised perceptibly every few months in the light of new problems or new data.

Vessels for toxic or explosive substances under the rules of A1 must have all welds radiographed except those at support brackets; this tends to restrict the types of weld that can be used. Under A2 and B any special-risk requirements are at the buyer's discretion to specify, for example, clause 3.4.1. in B.

As in other fields, it is becoming feasible to design with two attitudes; a general stress level for normal loads and higher level, into the yield region, for rare conditions such as initial assembly, hurricane or earthquake or some foreseeable once-only malfunction. This is not dangerous if we use ductile materials for vessels. Where cast iron vessels are used, the code rules are more stringent than for ductile materials.

The first and most vital part of the codes is the table of permitted materials with the permitted design stresses at various temperatures. Code B even considers creep by giving various stress levels according to the expected length of life.

The permitted working stress is not necessarily the same as the stress shown in the tables; in some circumstances a factor is applied; for example, in A2, 4–141, the calculated working stress can be two or three times the permitted tabulated stress.

The codes vary slightly over test pressures. A1 requires a test to 1½ times working stress, A2 and B to 1¼ times only. The test pressure is calculated to allow for elevated service termperature since tests at high temperature are rarely feasible; a further adjustment is made if the vessel has a corrosion allowance (extra wall thickness, often 1.5 mm per side). A1 has an extra facility for vessels not complying with the code design procedures. To get such a vessel stamped as safe it must pass a test to 2½ times working stress. This little-known provision is found in clause UG-101.

9.2 WALL THICKNESS REQUIREMENTS AGAINST INTERNAL PRESSURE

9.2.1 Cylinders and Spheres

The formulae are essentially equation 6.13c and its spherical equivalent, simplified for moderate D/d values and recast in terms of inner diameter only, together with rarely-needed versions when there are large wall forces in addition to the pressure forces. The terms are defined in figure 9.1. The values below are from code A2. The others tend to differ slightly.

Cylinder

$$T = \frac{pR}{(\sigma_d - \frac{1}{2}p)} \tag{9.1}$$

if $p > 0.4\sigma_d$

$$T = R(e^{p/\sigma_d} - 1) \tag{9.2}$$

Sphere

$$T = \frac{pR}{2(\sigma_d - \frac{1}{4}p)} \tag{9.3}$$

if $p > 0.4\sigma_d$

$$T = R(e^{p/2\sigma_d} - 1) \tag{9.4}$$

σ_d is the permitted design stress from the code tables, modified for fatigue if necessary.

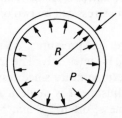

Figure 9.1

In cylinders the thickness is governed by the hoop stress, as explained in section 6.3. The axial stress is half this value. This point matters since we may sometimes use less efficient weld forms in round seams than in lengthwise seams; also we need less reinforcement around holes, in that direction. In spheres the stress is distributed symmetrically.

9.2.2 Cones and Cone–Cylinder Junctions

Theory: the pressure stress in a cone is as for a cylinder of the same true radius of curvature; this is $R \sec \theta$ in the notation of figure 9.2. The troublesome part is the radial component of the longitudinal forces; this amounts to $\frac{1}{2}pR \tan \theta$. This inevitably affects the hoop stress in the cylinder, increasing it at the small

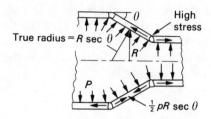

Figure 9.2

end; at the same time the flexure associated with it produces an extra axial tensile stress. At the large end the hoop stress in the cylinder is decreased but the flexural stress still occurs. The situation is similar to a flanged joint, section 9.5. For design purposes we simply use the expressions and graphs in the codes which are reasonably straight-forward.

9.2.3 End Covers

At first sight the logical end cover for a cylindrical vessel is a hemisphere. However, hemispheres are relatively expensive to make; moreover we note that the required thickness is only half that of the cylinder for the same pressure. This is inconvenient for making a good full penetration butt weld. Also, for a given stored volume a hemisphere takes up more length. The great majority of vessels use dished ends.

Typical dished end covers are ellipsoids or double-radius (torispherical) forms, with a short cylindrical skirt to get the weld away from the curved part. They are normally obtained from specialist manufacturers who have a stock of punch shapes and sizes to fit most needs. The method is pressing or spinning a circular plate over the punch, with free choice of thickness. The excess is machined off, leaving a suitable weld preparation shape. The designer tries to select a shape of dome *radius* slightly less than the vessel *diameter*, with a knuckle radius $\approx 0.1\ D$, or an ellipsoid of height $\approx 0.25\ D$ (plus the skirt length), then he calculates the required thickness as below.

The effect of internal pressure makes the body tend towards the shape of greatest volume/surface ratio, that is, the spherical. The flexure is most severe at the knuckle; moreover it combines with the inward pull towards the spherical outline. In thin vessels creases can form at this point. These are often harmless but under repeated loading have been known to cause fatigue cracking. A study of Adachi and Benicek [51] suggests that if $T > D/300$ there is no risk of creasing before reaching the permitted pressure.

A very satisfactory design procedure is given in B. From catalogues we choose a form with dome radius just less than the vessel diameter, with knuckle radius preferably $0.1D$ or so, or of ellipsoidal shape of height 0.2 to $0.25D$. Then we calculate the equivalent height h. It is the smallest of these three – actual height, $D^2/4(R + T)$ or $\sqrt{[(r + T)D/2]}$. Obviously this height excludes the cylindrical skirt (figure 9.3). This height is then used in a graph which plots the thickness required to satisfy the code in the form T/D against p/σ_d and a range of effective heights/D.

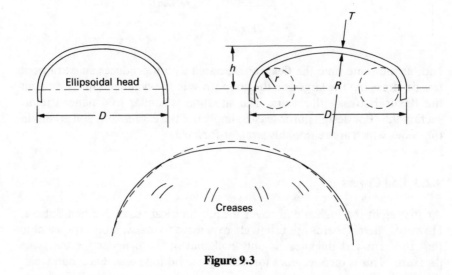

Figure 9.3

For a quick check, the thickness may be taken from $T/D = 0.0012 + [0.087p(D/h)^{1\frac{1}{4}}]/\sigma_d$ if $0.15 < h/D < 0.3$. This covers the great majority of standard shapes and gives answers within 5 per cent of the chart in B.

The real stresses are another matter altogether. There are analytical procedures for calculating the stresses in a given shape rather than for finding a shape which has low stresses. Photoelastic data disagree seriously with strain-gauge data on steel components. Reference 52, using strain gauges on a series of realistic 1 metre diameter steel heads, finds maximum stress concentration factors of $0.7\, r/T$, in the axial plane, at the inner surface of the knuckle, based on the hoop stress in the cylindrical part, $pD/2T$. However, we must not be tempted to use this as a sole guide; if we made the corner sharp in order to reduce r/T we should get a much worse stress concentration factor, theoretically about $1\frac{1}{2}\sqrt{(R/T)}$ (notice R, not r, in this expression).

When r is larger than $0.15D$, the conditions get closer to the spherical and the SCF is bound to get smaller. We must accept the idea that the elastic stresses are high and some yielding is expected.

9.3 EXTERNAL PRESSURE

External pressure is a common situation in tubular heat exchangers, in process vessels fitted with high-pressure steam heating jackets and in vacuum vessels. Less obvious cases are storage tanks and food processing vessels. Storage tanks develop partial vacuum during emptying unless they are open or unless vacuum relief valves are used. It is possible for such valves to fail, or the inert gas supply as used in some oil tanks may malfunction. The tank may be pressurised sufficiently by the vapour pressure of the contents until the temperature falls, then it could collapse. In food vessels this could happen after steam-sterilising, or generally after shut-down. Often it is economically feasible to make the design safe against the eventuality of 1 atmosphere net external pressure.

A cylindrical shell under external pressure is analogous to a strut; any deflection from the true circle causes bending which in turn aggravates the deflection. The force to cause buckling can be calculated on a strut basis by taking an equivalent length. In pressure vessels and pipes we often have considerable support from flanges, ends or deliberately designed stiffeners. If the unsupported length is such that $L/D > \sqrt{(D/T)}$ then the pressure at which buckling is to be expected is $\approx 2.2ET^3/D^3$. For shorter unsupported lengths L the pressure is $2.6E\,(T/D)^{2.5}(D/L)^{1.06}$ over a very wide range, namely $L/D\,0.3$ to 30, D/T 20 to 600. In dome-ended vessels the length is taken as the cylindrical length $+\ 0.4 \times$ dome height; this allows for average end conditions. In the codes these values are presented as design charts.

A spherical shell is of course much stronger. By the Karman–Tsien method published in 1939 the pressure at which a sphere is expected to buckle elastically is $0.33E'\,T^2/R^2$. A modified theory by Tsien [53] gives pressures in the region of $0.15\,ET^2/R^2$, supported by several experimental values. For some reason this work is little-known, possibly because it was published in war-time (1943). Bickell and Ruiz, [54] on p. 526, draw attention to buckling failures arising at pressures well below the 1939 Karman–Tsien value, closer to the 1943 Tsien value.

Unfortunately some existing books still refer to a much higher buckling stress value, attributed to Timoshenko. This is a kind of upper bound which might be approached by exceptionally true spheres but must be considered grossly unsafe in common practice, by a factor of 8 or more.

A2 gives a particularly convenient set of charts for calculating the permissible external pressure for cylinders and spheres. The author has compared a number of cases by the chart method and by elastic and plastic–elastic basic treatment and finds the chart to be on the safe side by a factor of about 4 for thinnish cylinders, 3 for thicker cylinders, 2½ for thin spheres and over 3 for thick spheres. The treatment in B is more elaborate and seems to have a safety factor of 5 for thin shells, 2 for thick ones; in the intermediate region where elastic and plastic failure are comparable, the factor for cylinders is down to 2, while for spheres if we accept Tsien, [53] A2's safety factor tends towards 1, in other words the safety margin tends to vanish in thin shells.

As the situation will mostly be that of a jacketed vessel, any buckled inner parts are not visible. On the whole the author greatly prefers the A2 method, on grounds of certainty as well as convenience. Since it requires thicker walls for a given duty, it gives designs which amply satisfy B.

9.4 NOZZLES

Nozzle is the term for a short connection welded to a pressure vessel. It may be subsequently welded to a pipe, or may carry a flange or a bellows unit. The shell is weakened by the presence of a hole but below a certain size the codes excuse us from reinforcement. This is based on experience and may need reviewing as higher stress levels come into use. In most cases some form of reinforcement is needed. Openings are restricted in size (see below); also we should keep them away from such features as welded seams, the knuckle region of dished ends and support points. We also keep them as far as possible from each other. If they must be closely grouped, then we reinforce the group as if it were a single large hole. The codes give detailed rules.

9.4.1 Theory

A small hole in a spherical vessel wall under pressure gives an SCF of 2 where the tension is symmetrical, or about 2½ in a cylinder where the hoop stress and axial stress are in 2:1 ratio. Flexibility of the shell reduces the stress concentration since the edge can cave inwards, thus reducing the periphery. To visualise this, figure 9.4 shows a vessel with a hole covered by a flexible patch. This patch is supported by an external pillar so that we could measure the behaviour of the shell by itself, unaffected by the closure method. Despite the relief due to flexing, we expect the hole edge to be under a tensile stress greater than the unpierced shell far away from the hole. Typical stresses are many times greater than the nominal shell stress; figure 9.4 shows SCF values

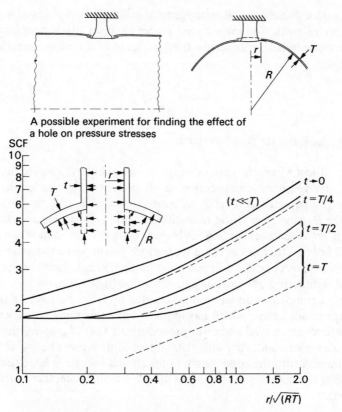

A possible experiment for finding the effect of
a hole on pressure stresses

Stress concentration factor for spherical shells with nozzles
Definition: SCF = max. stress/($pR/2T$)
Solid lines apply to flush nozzles, dashed lines to
inward-protruding nozzles

Figure 9.4

from Leckie and Penny, [56] replotted on log–log paper. Values for cylinders
are more complicated, generally somewhat higher.

When a nozzle is welded into or on to the shell, the rim is no longer entirely
free to relax. The main effect is just the weakening due to the hole, like in a
flat plate in tension. One possible remedy is to make the whole shell and the
nozzle extra thick, reducing the basic stress so that even with the stress
concentration we do not exceed an acceptable stress level. This is done when a
vessel has a great many outlets like a steam drum.

In other cases it is more economical to reinforce locally. Reinforcement
may consist of any mixture of the following: special forged joint-pieces,

nozzles with a thickened end, extra general wall thickness of vessel or nozzle, or reinforcing pads. The forged joint pieces are particularly beneficial if full X-ray inspection is required since they are set in with butt welds rather than fillet welds.

9.4.2 Design Rules for Reinforcement

This is a subject where the codes diverge considerably. A1 requires a putting-back of the removed cross-section in all directions, for holes up to the permitted maximum of $d = 0.3D$. A2 is more permissive, it allows the same up to $d = 0.5D$; moreover if the transition is made very progressive the compensating cross-section supplied can be ¾ of the lost cross-section or even less for small holes. The rules in B are more closely defined and somewhat more demanding than A1 or A2. Certain small inspection holes are exempt and do not need reinforcing, relying on edge relaxation (figure 9.5, top).

The area compensation method is accepted by B but only where previous good experience exists, which has implications concerning stress level. B's main method uses a trial and error procedure on a set of graphs; this is quite quick once understood. An outline of the procedure is given below, to be used in conjunction with the code. Some of the ifs and buts have been simplified. The graphs are not shown, to prevent misuse due to incomplete instructions. Notation:

σ = design stress, p = design pressure;

D, d are shell and branch diameters respectively without corrosion allowance;

T, t are shell and branch thicknesses derived from the pressure and stress far away from the hole;

T_r, t_r are the shell and branch thicknesses in the vicinity of the hole as required by the rules. These are what we are trying to find. These also are calculated before corrosion allowance is added on.

Procedure:

Find the duty of the vessel: pressure, temperature, design life, pressure cycles for fatigue assessment, external forces, corrosion allowance required (this may slightly affect D and d), proximity of other openings, special material requirements if any.

Select material provisionally unless specified above;

Find design stress from tables 2.3 in code;

Check for fatigue (see section 9.6), amend design stress if necessary;

Check whether d/D is within design rules (0.33 maximum for cylinders, 0.5 maximum for spheres). If nozzle is oblique, check with clause 3.5.4.3.6

since the effective size depends on whether the hole is biggest in hoop stress direction or the other way.

Find T, t from pressure formula (equation 9.1 for cylinder, $T = pD/(4\sigma - 1.2p)$ for sphere) (marginally different from A2 version).

Look into extraneous complications such as pipework or support forces, bending of horizontal vessels, etc., (code appendices B, G). If these give stresses in tension plus bending or compression plus bending $< \frac{1}{2}\sigma$, forget them. If they are above this, follow code appendix B. This takes a great deal of concentration.

Look up code table 3.5.4(1) to find the smallest allowable nozzle thickness for the specified d value. If greater than previous t, use new value from now on.

From here on it is easier to consider a numerical example.

Example 9.1

Let $D = 450$ mm, $d = 100$ mm, $P = 50$ bar $= 5$ N/mm^2, $\sigma = 100$ N/mm^2. The nozzle is sufficiently far away from other nozzles, etc., not to require a modified treatment. Extraneous stresses are below the limit. The vessel is cylindrical. There is no fatigue.

$$T = \frac{5 \times 450}{200 - 5} = 11.54 \text{ mm}$$

$$t = \frac{5 \times 100}{200 - 5} = 2.56 \text{ mm}$$

but by table 3.5.4(1) $t = 5.4$ mm.

Find out whether nozzle may protrude inside. If so, use graph of figure 3.5.4(1). We assume that protrusion is not allowed since this brings out an important aspect of the procedure. Hence use figure 3.5.4(3) for flush nozzles in cylinders up to $d/D = 0.3$ and also supplementary graph figure 3.5.4(4) for d/D above 0.2, as follows.

From common sense we must expect to finish up with $T_r > T$. Try $T_r = 1.5T = 17.3$ mm. Calculate the flexibility parameter

$$\rho = \frac{d}{D}\left(\frac{D}{2T_r}\right)^{\frac{1}{2}}$$

If you prefer, you can cancel this down to

$$\rho = \frac{d}{\sqrt{(2DT_r)}}$$

In theory these should be mean diameters $d + t_r$, $D + T_r$, but it won't make much difference, so we forget about it

$$\rho = 0.826$$

$$\frac{t_r}{T_r} = \frac{5.4}{17.3} = 0.312$$

From graph

$$\frac{CT_r}{T} = 1.55$$

C is a slight correction factor for pipe stress if no major stresses were included above. For typical well laid-out work, $C = 1$. If external loads are prevented by bellows, etc., $C = 1.1$. We take $C = 1$. So we seem to be nearly there.

$$\text{New } \rho = \frac{1.017}{\sqrt{1.55}} = 0.817 \qquad \frac{t_r}{T_r} = 0.303$$

Ans. $T_r = 1.55T = 17.9$ mm

(If we cycle again with new ρ, we get virtually the same answer.)

Now since $d/D > 0.2$ we must use the supplementary graph also. Find $Y = (t_r/d)(2D/T_r)^{\frac{1}{2}}$. Using the last T_r

$$Y = \frac{5.4}{105}\left(\frac{911}{17.9}\right)^{\frac{1}{2}} = 0.367$$

From the graph, $CT_r/T = 1.77$; C is still $= 1$, $T_r = 20.4$ mm.
Recycle with new T_r; $Y = 0.343$, $CT_r/T = 1.83$, $T_r = 21.1$ mm.
If we cycle again, Y will get smaller, the required T_r will be larger still, then Y gets still smaller; the answers diverge more and more.
This is understandable if we take the hint from figure 9.4. Although this relates to spheres rather than cylinders, it does show what high stresses result when $t \ll T$, compared with the basic shell stresses. To reach a converging answer we must use the supplementary graph in reverse, although the code does not warn us about this. By taking T_r/T (or CT_r/T if $C \neq 1$) from the previous calculation and entering the abscissa, the answer is found on the ordinate. In this example, $Y = 0.5$. Substituting this in the expression defining Y, $t_r = 105 \times 0.5 \ (17.9/936)^2 = 7.26$ mm. By clause 3.4.5.3.3 we may interpolate between the old and new t_r, giving (in this case) 6.33 mm for the final t_r.

How far must the new thicknesses T_r and t_r extend?
The essential region of extra thickness is defined in figure 9.5, top view. From this we can work out how much cross-section is being provided, purely for interest. The design is satisfied already. In this instance it is 1260 mm². By area compensation we would need $100 \times 11.5 = 1230$ mm², and we would not realise that some of this should go into the nozzle.
On the placing of reinforcement the codes agree in general principle but

diverge widely in detail, particularly in the amount we can take credit for in the nozzle length direction. To save space and confusion we only show the alternatives from B, the upper to go with the graph procedure just outlined and the lower with the full area replacement rule. There are a number of restrictions to ensure that the material is distributed for good elastic matching, without overloading the fillet welds where pads are attached. The material used for pads must be strong enough for the design stress; yet if it is stronger it must not be made thinner because its elastic contribution is just as vital as its strength (figure 9.5).

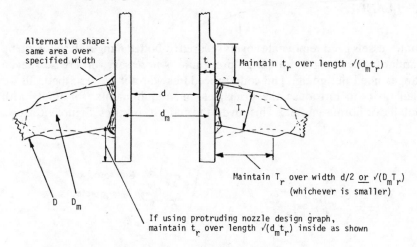

Alternative shape: same area over specified width

Maintain t_r over length $\sqrt{(d_m t_r)}$

Maintain T_r over width $d/2$ $\underline{or}$ $\sqrt{(D_m T_r)}$ (whichever is smaller)

If using protruding nozzle design graph, maintain t_r over length $\sqrt{(d_m t_r)}$ inside as shown

Minimum reinforcement to BS 5500 Issue 4 clause 3.5.4.3.4

Reinforcement can be forged plug or pads and fillet welds. Transitions as smooth and gradual as practicable.

Welds are indicated schematically. See sample designs in code.

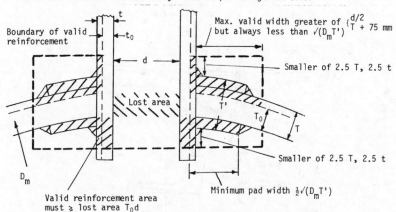

Boundary of valid reinforcement

Max. valid width greater of $\{\begin{smallmatrix} d/2 \\ T + 75 \text{ mm} \end{smallmatrix}$ but always less than $\sqrt{(D_m T')}$

Smaller of 2.5 T, 2.5 t

Lost area

Smaller of 2.5 T, 2.5 t

Minimum pad width $\frac{1}{2}\sqrt{(D_m T')}$

Valid reinforcement area must $\geqslant$ lost area $T_0 d$

Definitions for compensation by area replacement to BS 5500, appendix F. Restricted to previous satisfactory experience (application and stress). Beware of notation variations in this part of the Standard.

Figure 9.5

Codes A1 and A2 have different variations on the theme of the required regions. There is scope here for a new rule to coordinate US and UK requirements.

Readers using BS 5500 appendix F are warned to study the notation relevant to the appendix; this uses T_r as the thickness of reinforcement added, not the required total thickness as in the main code.

9.5 FLANGES

Some vessels need removable covers secured by bolted flanges. If these can be standard pipe flanges, no design procedure is necessary; otherwise special flanges need designing. The code procedures will not be described, all we shall do is to introduce the principles and requirements, for joints with metallic or fibrous gaskets. This covers the majority of cases. Figure 9.6 shows

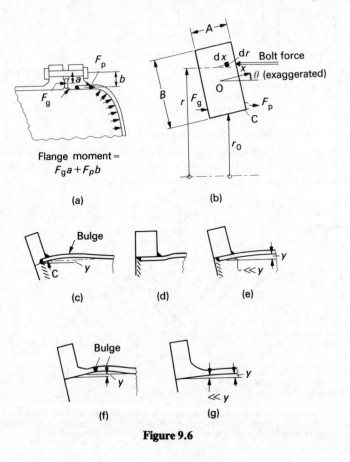

Figure 9.6

two regular flange types and a typical wide, flat gasket lying within the bolt circle. A great variety of gasket materials and seating arrangements are used. A gasket needs some plasticity to take up initial mismatches, fill up depressions or scratches and accommodate the distortion due to bolting-up. After that it needs elasticity to maintain a seal when the pressure forces cause some deflections. To maintain a leakproof joint we compress the gasket to a pressure two or more times the fluid pressure being sealed. The width and thickness of the gasket are dictated by experience. Seating pressures are suggested in the code for a great range of materials.

O rings and lip-seals are self-energising and require negligible initial pressure. The flanges for these are bolted face-to-face, with slightly different design rules.

In the bolting-up condition we have a moment equal to the seating force times the distance from this to the bolt circle. In the working condition the pressure in all cases (except with certain expansion joint arrangements) causes an axial pull in the wall, plus a further force on the clear face between inside diameter and gasket seat. This is best taken into account as (fluid pressure × area up to the gasket). The line of action is usually about the middle of the shell thickness. This force and the gasket force times their distances from the bolt circle make up the working moment, usually greater than the bolting-up moment. Cases where the working moment is smaller are vacuum vessels and some cases of thermal expansion or contraction.

The flange moment twists the flange, extending one face and compressing the other face down to a smaller radius. The analysis for a plain flange is relatively simple. In figure 9.6b, the axial thickness = A, the radial breadth or width = B and the mid-base radius r_0 does not change. Note that BS 5500 uses B for the inner diameter (bore); we prefer to reserve the letters A, B for widths.

An elemental ring, position x, r, of dimensions dx, dr is extended from radius r to radius $r + x\theta$ due to the twist θ. It is under stress $E \times x\theta/r$, that is, $E \times$ extension/length. Proceeding by energy, the strain energy per unit volume = stress$^2/2E$.

$$U = \iint \frac{E^2 x^2 \theta^2}{r^2} \times \frac{2\pi r}{2E} \, dx \, dr \quad \text{limits } r_0, \frac{-A}{2} \text{ to } r_0 + B, \frac{A}{2}$$

$$U = E^2 \theta^2 \frac{A^3}{12} \times \frac{2\pi \ln \left(1 + \dfrac{B}{r_0}\right)}{2E} \tag{9.5}$$

But a moment M causing a twist θ does work = $M\theta/2$, hence $U = M\theta/2$

$$M = \frac{\pi}{6} E\theta A^3 \ln \left(1 + \frac{B}{r_0}\right) \tag{9.6}$$

The highest stress

$$\sigma = \frac{3M}{\pi A^2 r_0} \left| \ln \; 1 + \left(\frac{B}{r_0} \right) \right.$$ (9.7)

Note how the energy approach removes any doubts about using the 2π factor correctly.

In large vessels $B/r_0 \ll 1$ so we can expand the logarithm into the series $B/r_0 - B^2/2r_0^2 \ldots$ and take only the first term

$$\sigma = \frac{3M}{\pi A^2 B}$$ (9.8)

Similarly for the moment equation 9.6

$$\frac{M}{\theta} = \frac{\pi E A^3 B}{6r}$$ (9.9)

We need a way of allowing for the bolt-holes. Bernhardt [55] shows that an equivalent width B' agrees with the experimental deflection stiffness where

$$B' = B - \frac{nd^2}{\pi D}$$ (9.10)

when the mean diameter is D and there are n holes of diameter d. This agrees with the author's analysis in section 7.13. There is little evidence on the local stresses so we may assume that figure 4.12 applies, but it is hardly relevant since these will largely disperse due to local yield during bolting-up.

Fillet-welded flanges are difficult to X-ray effectively so for some applications we must use taper-hub flanges which extend some distance before coming down to shell thickness. These flanges must not be cut from plate with the grain parallel to the face because the thin neck would be under tension in a direction in which plate material is weakest.

The problem with flange design is the effect it has on the shell. The radial growth of the flange as a whole is small compared with the shell under internal pressure, so the flange acts like a corset. If we designed the flange to a maximum stress equal to the vessel hoop stress, then corner C, figure 9.6c would expand as much as the shell; owing to the tilt, some parts of the shell would be overstretched and overstressed tangentially, that is, to an excess hoop stress. As we shall see later there is also a substantial axial stress due to combined bending and tensile force. An undeflected flange (d) is neither feasible nor beneficial. Arrangement (f) is more progressive in flexure but still has the bulge. The best compromise is (e) or (g), with stress and strain at the flange corner kept well below the shell hoop stress and strain.

The presence of a tapered transition does not remove the basic arguments; it merely makes it difficult to pinpoint the exact spot where the flange stops and the shell starts to flex. We still get a very fair idea of the bending moments and stresses by using the analysis given below. Bernhardt [55] agrees that if the flange corner is stressed to the general design hoop stress then the shell will be overstressed but he considers this acceptable in pipe-work. We note that in the code design rules the flange corner is allowed to go well above the design stress, so the shell stress would rise further still. The author considers this very unwise when the high stress points include regions such as welds or the heat-affected zone near to welds. The following analysis shows that these high stresses can be avoided by thickening the flange moderately.

This is not an academic exercise; flange yielding and leaking are a serious hazard (a) with combustible products, (b) when hidden by lagging and in-accessible for retightening and (c) with toothed locking rings for removable covers. The author is aware of a fatal accident associated with a flange designed by code which yielded and allowed a cover to blow off.

The analysis is of second/third year standard. The notation is shown in figure 9.7a. The base-line condition is the shell in the working state. The

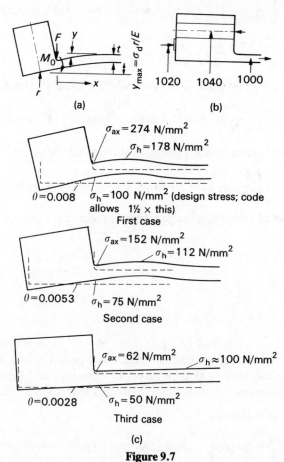

(c)

Figure 9.7

corseting effect of the flange is given in terms of an inward force per unit periphery, and an inward moment also per unit periphery. The deflection is positive inwards, likewise the slope. This makes the case consistent with the usual beam-on-elastic-foundation notation. The elastic foundation is due to the compression of the shell – any inward deflection y causes a hoop stress change Ey/r. This is equivalent to an outward pressure by analogy with the pressure-stress relation in a thin cylinder, since a pressure p would cause a hoop stress pr/t, thus the stress change Ey/r is equivalent to a pressure $Ey/r \div r/t = Eyt/r^2$.

For the beam-like action per unit periphery

$$E'I\frac{d^4y}{dx^4} = p = \frac{Eyt}{r^2} \tag{9.11}$$

The solution (see appendix A.8) is

$$y = \left(\frac{e^{-mx}}{2m^3E'I}\right)(F\cos mx - mM_0\cos mx + mM_0\sin mx) \tag{9.12}$$

$$m = \left[\frac{3(1 - \nu^2)}{r^2t^2}\right]^{1/4} \approx \frac{1.29}{\sqrt{(rt)}} \tag{9.13}$$

We shall require the end slope of the shell where it joins the flange

$$\frac{dy}{dx} = \frac{e^{-mx}}{2m^2E'I}\ (2mM_0\cos mx - F\sin mx - F\cos mx) \tag{9.14}$$

For checking the highest bending stress we also require the second differential

$$\frac{d^2y}{dx^2} = \left(\frac{e^{-mx}}{2mE'I}\right)(2F\sin mx - 2mM_0\sin mx - 2mM_0\cos mx)$$

The clearest way of presenting the point of this method is a trial design (figure 9.7b, c).

Example 9.2

Consider a steel vessel, radius 1 m, pressure 1 N/mm², design stress $\sigma = 100$ N/mm². The vessel is to have a bolted flange with a gasket 20 mm wide, mean radius 1020 mm and is to work under a contact pressure of 3 N/mm². There will be n bolts of diameter d (to be found), stressed to 180 N/mm² on root area.

   (1) Find wall thickness

$$t \approx \frac{pr}{\sigma} = 10\ mm \quad \text{(should be 10.05 mm!)}$$

(2) Find forces

$$\text{Pressure force} = p\pi r^2 = 3.14\,\text{MN}$$

$$\text{gasket force} = 3 \times 2\pi \times 1020 \times 20 = 385\,\text{kN}$$

(3) Provisional bolt selection

$$\text{total bolt area required} = (3.14 + 0.385) \times \frac{10^6}{180}\,\text{mm}^2$$

Number of bolts depends on spacing; this can hardly be less than $4d$. Pitch circle roughly 1040 mm radius, hence

$$n \approx \frac{2\pi \times 1040}{4d}$$

$$\text{Total area} = \frac{n\pi d^2}{4} = \left(\frac{2 \times 1060}{4d}\right)\left(\frac{\pi d^2}{4}\right)$$

Ans. $d = 15$ mm (root), $n = 110$.

This will be modified to suit standard thread and conventional number (multiple of 4) but is sufficient for now, so that we can decide on a flange width B;

(4) Find flange moments provisionally

$$\text{pressure moment} = 1 \times \pi \times 1010^2 \times 30\,\text{N mm}$$

$$\text{gasket moment} = 0.385 \times 20 \times 10^6\,\text{N mm}$$

$$\text{total} = 104 \times 10^6\,\text{N mm}$$

(5) Find equivalent width B': B can be ≈ 60 mm; B' can be found by equation 9.10. Bolt *holes* say 18 mm

$$B' = 60 - \frac{110 \times 18^2}{2\pi \times 1040} = 55\,\text{mm}$$

However, to give latitude and rigidity, let $B = 85$ mm, $B' = 80$ mm, using a larger bolt circle, total moment 130 M N mm;

(6) Find A for three alternative stresses at corner: 100, 75, 50 N/mm². From equation 9.8

$$A = 124, 143, 176\,\text{mm}$$

(7) Find θ: from equation 9.9 or from strain, $\theta = 0.00814, 0.0053, 0.0028$ radians;

(8) Find y: this is the radial expansion of the vessel less the radial expansion at the flange corner. $y = 0, 25r/E, 50r/E$, that is, $y = 0, 0.125, 0.25$ mm;

(9) Investigate stresses in shell: at $x = 0$

$$y = \frac{(F - mM_0)}{2m^3E'I} \quad \text{from equation 9.12}$$

$$\theta = (2mM_0 - F)/2m^2E'I \quad \text{from equation 9.14}$$

Take first case:
$y = 0$, therefore $F = mM_0$, $\theta = 0.008$
therefore $mM_0 = 0.008 \times 2\, m^2E'I$ $\quad M_0 = 3733\,\text{N/mm}^2$

$$\text{Bending stress} = \frac{6M_0}{t^2} = 224\,\text{N/mm}^2 \quad \text{at } x = 0 \text{ (maximum)}$$

$$\text{axial tensile stress due to pressure} = \tfrac{1}{2}\sigma_d = 50\,\text{N/mm}^2$$

$$\text{total tensile stress} = 274\,\text{N/mm}^2 \quad \text{axial, at flange corner}$$

For maximum hoop stress we need greatest negative y; found from $dy/dx = 0$.
Hoop stress $= 100\,\text{N/mm}^2$ plus extra due to negative y_{max} at bulge. From
equations 9.12 and 9.14, the maximum occurs when $mx = \pi/4$.
Maximum bulge $= 0.39$ mm, extra hoop stress $= 0.39E/r = 78\,\text{N/mm}^2$, at $x =$
61 mm; $\sigma_h = 178\,\text{N/mm}^2$.
Take second case:
$y = 0.125$ mm, $\theta = 0.0053$. From equation 9.12

$$F - mM_0 = 0.125 \times 2 \times \frac{1.29^3E'I}{(rt)^{1.5}} = 9.72$$

From equation 9.14

$$2mM_0 - F = -0.0053 \times 2 \times \frac{1.29^2E'I}{rt} = 31.9$$

$mM_0 = 22.2$, $M_0 = 1720$ N mm, bending stress $= 103\,\text{N/mm}^2$, total tensile
stress $= 152\,\text{N/mm}^2$, maximum reverse $y \approx 0.06$ mm, extra tensile stress $= 12$
N/mm^2, at $x = 93$ mm.
Third case:
$y = 0.25$ mm, $\theta = 0.0028$.

$$F - mM_0 = 0.25 \times 2 \times 1.29^3 \times \frac{E'I}{(rt)^{1.5}} = 19.4$$

$$2mM_0 - F = -0.0028 \times 2 \times 1.29^2 \times \frac{E'I}{rt} = -16.9$$

$mM_0 = 2.5$, $M_0 = 194$ N mm/mm, bending stress $= 6M_0/t^2 = 11.6\,\text{N/mm}^2$; σ_{ax}
$= 62\,\text{N/mm}^2$. Bulging extra stress is negligible.
    Intermediate case $y = 0.2$ mm, $\theta = 0.0037$. This gives bending stress 32
N/mm^2. Bulging extra stress 3 N/mm^2. This calculation errs slightly on the

pessimistic, conservative (safe) side since it neglects radial increase of the flange as a whole.

To conclude, we see that by designing flanges to a stress of $0.6\sigma_d$ which calls for a one-third increase in flange material, the shell stresses can be greatly reduced. This seems to hold true for other thicknesses and diameters too.

For taper-hub flanges the argument is not very different since the main effect is a rearrangement of material. The bending stress is still liable to be high near the weld if the flange is designed to reach σ_d. The detailed design of taper-hub flanges is laid out in codes A2 and B. B is particularly helpful in giving a specimen worksheet. The following calculation was made according to code procedure but retaining the 100 N/mm² design stress.

Adding a small taper-hub to the flange used above, (case 1), the hub being 15 mm high axially, tapering from 15 mm to 10 mm, gives a hoop stress of 101.4 N/mm² and an axial stress of 167 N/mm², in the hub. It is not clear from the code whether this is the flexure stress only or whether it includes the axial tension. Moreover there is no provision for finding the consequential stress in the shell due to bulging.

9.6 FATIGUE ASSESSMENT OF PRESSURE VESSELS

Pressure vessels are usually in a good position to resist high cyclic stresses. Firstly the loading is nearly always unidirectional. In a highly ductile material, notably in mild steel, the one-way endurance limit is close to the yield point. The pressure test, particularly if taken to 1½ times working stress, puts in favourable residual stresses.

The routines in codes A2 and B agree in principle and in many details. The basis is a permitted range or a permitted amplitude. The stress is specified as the stress intensity, defined as twice the highest shear stress. Since this is usually an out-of-plane shear, the criterion is nearly always simply the highest tensile stress.

The first step in fatigue assessment is a procedure to see whether we are exempt from the problem. The procedure looks rather confusing in B (A2 does not specify an exemption procedure). The following is a plan for checking on exemption according to B. The data worked out will then serve readily for fatigue assessment if needed.

(1) Check whether design contains partpenetration fillet welds, screwed connections or heat exchanger tube sheets. If so, assume that the *life* is halved for a given *stress*. Either double all the actual number of stress cycles or sketch in a supplementary curve shifted to the left on the graph provided. Don't get worried by the high stress values on the scale; these are notional values and get modified later.

(2a) Take the design stress for the material and multiply by 3.

(2b) Find the number of cycles on the graph corresponding to this strange value. If the material is not steel, modify stress in ratio of E, or draw an alternative curve for your material, for example, for aluminium the stresses need to be about one-third of the steel stresses at the same number of cycles.

(2c) If the number of expected full range cycles including start and stop is less than the number found in 2b, go to 3a.

(3a) From 2b find the permissible stress *amplitude* for 2 000 000 cycles using the 2 to 1 adjustment from stage 1 if necessary.

(3b) Divide by 3.

(3c) Find the *range* of stress due to pressure cycles expected in service. If less than the 3b value go to 4a.

(4a) Find the highest cyclic temperature difference θ expected between two points less than $2\sqrt{(RT)}$ apart. With the coefficient of linear expansion α, calculate the stress range doubled, $2\alpha\theta/E$.

(4b) From the graph modified as necessary (see 2b) find the stress figure for the number of start-up and shut-down cycles.

(4c) If $2\alpha\theta/E$ is less than this, go to 4d.

(4d) Find the number of temperature cycles during running (not start and stop).

(4e) If the stress value $2\alpha\theta/E$ is less than the permissible value from the graph for this number, go to 5.

(5) Find the *range* of stresses due to mechanical forces and the number of such cycles expected. If below the value given by the appropriate curve, go to 6.

(6) Rejoice; there is no need to assess for fatigue.

If you have got stuck at any stage along this route then fatigue assessment is needed. The data obtained come in useful using Miner's law (section 3.4.3). It may be possible to redesign some details, notably if the culprit is stage 4a. A typical remedy is given below.

For bolts the situation is confusing. In the text, stresses in bolts can be up to $2S_m$ in tension, $2.7S_m$ if bending is allowed for. On the graph, a note allows stresses up to $3S_m$. In code A2, S_m is the listed design stess. Code B has accepted the same values, but fails to define S_m. If we take S_m to have the same meaning as in A2, namely the listed design stress, then we have the paradoxical situation that $3S_m$ is considerably above the UTS for several of the listed materials.

We may be astonished to note the numerical values on the design stress curves. At the low stress long-life end the values are well below the endurance limit, yet at 100 cycles the values exceed the UTS. The design values are of course limited by the overriding values given in the tables of materials, which lie between 0.25 and 0.4 of the UTS. The reason for the high values is that the graph provides a way of accounting for the washing-out of stress concentrations during the pressure test (see section 3.4). Compared with the *s–N* curve approach, the code seems safe by a factor of 4 at the high-cycle end (factor of 2 or more on stress, times the hidden factor since we calculate stress *range* but use the curve for *amplitude*). At the low-cycle end it seems rather less safe especially if stresses are expected to go into reverse (see appendix A.5 for further discussion).

In this connection it is worth pointing out that stress concentrations are ignored in the steady stress calculation but must be included in fatigue calculations. The usual values should be used as in chapter 4. For fillet welds a factor of 4 is required, based on the customary assumption that nominal stress = force/throat area. At the toe of a fillet weld, B is content with SCF = 2.5.

A likely point for high thermal stress cycles is the inlet to a thick vessel working an intermittent process. Hot fluid being fed into a cold vessel, or even worse, cold fluid into a warm vessel causes local expansion or contraction to happen much more quickly than the thick main body can hope to follow. The calculation requires data on the local heat transfer coefficient, the thermal conductivity of the material and the relative thicknesses. The effect is more severe with liquids than with gases. A design method to minimise this is the thermal sleeve, figure 9.8. A similar idea is used for mounting bearings on shafts which get hot intermittently or get too hot for the bearing. By making the heat path long and much thinner than the main body it acts as a protective resistance; the temperature gradient in the thick body is greatly reduced; in the configuration shown the reduction could be 100 to 1 compared with a direct connection if the annulus is packed or shrouded in some way. It is important to note that BS 5500 clause C 6.4 states that pad-type reinforcement is unsuitable where thermal stress cycles are significant. This seems to be a mandatory statement; if so it has tremendous significance not only for nozzle reinforcement as in section 9.4 but also for supports (section 9.8).

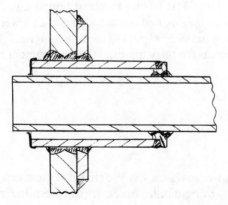
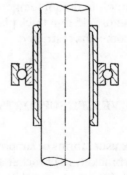

Figure 9.8

9.7 BRITTLE FRACTURE ASSESSMENT

Brittleness and toughness were discussed to some extent in section 3.5. Pressure vessels are generally not under impact; brittle fracture is rare but owing to the grave consequences it must be guarded against. Since, happily, there is little experience of brittle fractures in service, the failures often occurring during the pressure test and traceable to defects, the code procedures are tentative. The mainstay is the requirement of certain impact energy values based on standard or sub-standard size Charpy specimens. At the customer's insistence all materials used, however good their general reputation and 'static' properties from low-speed tests, may be required to show adequate energy absorption in a Charpy test. The codes lay down ways of scaling the acceptable toughness figures when under-sized specimens are used and also allow many materials to be exempt from tests.

In appendix A.4 we claim that even the standard Charpy specimen is too small to consistently predict the properties of thick material, let alone a sub-standard size specimen.

As mentioned earlier, brittleness at low temperatures is a feature of ferritic steels. Charpy specimens indicate a rate of change of toughness with temperature which seems to make it possible to assess toughness at one temperature from tests at a higher temperature. In appendix A.4 the author tries to cast some doubt on this proposition. The modest cost of larger specimens and chilling seems a small price for additional safety, avoidance of costly collapses on test and possible in-service failure.

The alternative method based on fracture toughness data also comes in for some searching into possible divergences between test and service conditions. Full-scale tests on pipelines have shown running cracks when theory suggested that conditions were safe.

Finally, attention is drawn to the changing views on safe design temperatures. The BS Draft for Development DD 55, 1978, differs appreciably from BS 5500:1976. In figure 6.11, p. 69 of DD 55 the levels of temperature at which certain Charpy values must be maintained or exceeded are set lower in relation to design temperature than in BS 5500 p. D18 for thin materials (for example, 15 mm thick plate), whereas for thick material the requirement has become less restrictive.

9.8 VESSEL SUPPORTS

The usual forms of support are shown in figure 9.9. We shall call them briefly (a) bracket, (b) bracket and pad, (c) rigid pillar bracket, (d) base pillar and pad, (e) ring mount, (f) saddle and (g) skirt.

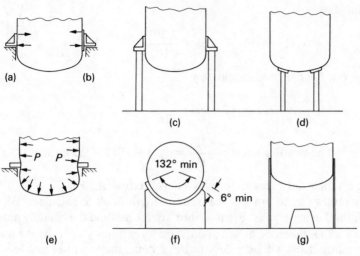

Figure 9.9

Theory

A support has two quite distinct effects on the vessel. The first and often the most serious is that it interferes with the normal expansion due to pressure; the second is the local flexure effect due to the load on the support. All the forms shown interfere with the expansion of the shell under internal pressure. Forms (e) and (g) only affect the axial stress which is half the hoop stress; the others affect the hoop stress in the same way as a class W or S detail in figure 8.12, amounting to a four-fold weakening if the pad thickness is of the same order as the shell thickness.

The ring support forms a cincture causing flexural stresses in the axial direction which are readily calculated using the treatment as for flanges. The ring would normally be well away from the vessel's ends, so we consider the slope under the ring to be zero. In equation 9.14 the initial slope at $x = 0$ is zero if $F = 2mM_0$. In most cases we can say that the ring is so thick that its radial expansion is negligible; thus the radial deflection y imposed on the shell is equal and opposite to the radial expansion due to the pressure

$$y = \sigma_h(1 - \tfrac{1}{2}\nu)\frac{r}{E}$$

where σ_h is the hoop stress due to the pressure. Note that we have allowed for the Poisson contraction due to axial shell stress; in section 9.5 this refinement was omitted for simplicity.

Now we substitute $F = 2mM$ in equation 9.12 at $x = 0$

$$y = \frac{mM_0}{2m^3 E'I}$$

and from above

$$y = \sigma_h(1 - \tfrac{1}{2}\nu)\frac{r}{E} \qquad m = \frac{1.29}{\sqrt{(tr)}}$$

Then we put $I = t^3/12$ and eliminate y

$$M_0 = 2m^2 E' t^3 \sigma_h (1 - \tfrac{1}{2}\nu)\frac{r}{12E}$$

$$\text{The bending stress} = \frac{6M_0}{t^2} = \sigma_h(1 - \nu^2)(1 - \tfrac{1}{2}\nu)m^2 tr = 1.3\,\sigma_h$$

Adding on the axial stress $\tfrac{1}{2}\sigma_h$ gives a tensile stress of $1.8\,\sigma_h$.

The stresses due to bracket loads are more difficult to calculate. BS 5500 has a method using graphs. In the author's quick method the bending moment per unit width is taken as inward or outward force times $\sqrt{(rt)}$, so for a force of F the bending stress will be $\approx 6F\sqrt{(rt)}/t^2$, F being the force per unit length of edge. This contains an allowance for concentration at the corners of a small pad. Some discretion must be used in deciding what the effective length of loading edge is when the bracket is backed by a load-spreading pad or when it has a central stiffening rib.

This expression gives results close to but safely above the code answers in a number of test calculations. The code unfortunately runs out of data when the tangential width of a support exceeds four times the axial length. Seeing that saddle supports should extend over 132° of arc, figure 9.9f, they are normally well above this ratio. The author's method is not subject to this difficulty.

Far more difficult and more serious is the case of pads with small subtended angles. As noted in section 8.6 which also deals with radial forces on cylindrical shells, when we approach flat plate conditions much higher flexure stresses can be expected; if the force is radial, section 7.12 seems appropriate. When the forces are due to the tilting moment M on a pad of square shape, size $a \times a$, on a vessel of diameter d, thickness t, the stress could be about $(3M \ln d/a)/at^2$ (derived from data in reference 5).

The best form of support for vertical vessels seems to be a thin skirt. This is free from hoopwise discontinuity; the corseting effect depends on relative thickness but is much less than with a thick ring; the weakness due to welding heat and weld shape acts only in the axial direction where the main stresses are smaller than hoopwise. The next best seems to be the rigid pillar bracket provided that the first pad welded to the shell is much thinner than the shell and/or tapered out. The rigidity is important so that the load on the shell is mainly shear, with as little flexural strain as possible.

Some points to bear in mind are

(1) For pressure testing, vessels are whenever possible filled with water; this reduces the stored energy in case of fracture. In thin vessels the weight of water can be an extra loading case when designing supports; in horizontal

vessels the weight of water can cause extra bending as marked at (a) in figure 9.10. This calls for wide saddles and may even require the use of intermediate plates to reduce stress concentration. An unusual alternative is to immerse the vessel in water for the test, then the external and internal *static* water pressure can be arranged to cancel out by matching the filling and emptying rates to maintain the same level inside and outside; the *test* pressure is then distributed uniformly as in a gas-filled vessel.

(2) A thin vessel may buckle axially, particularly when full of liquid but not under pressure. The supports should be placed at optimised points (section 7.6). As a guide to this danger, a cylinder may buckle in this way when the axial stress is $1.2Et/d$ in a perfect cylinder, and at much lower stresses if the cylinder is dented. A dent of $0.5t$ can halve the buckling strength.

(3) The pillars may buckle as short columns especially if the vessel sways due to wind, pipe forces, thermal expansion inequalities or ground settlement. Triangulation is a possible precaution.

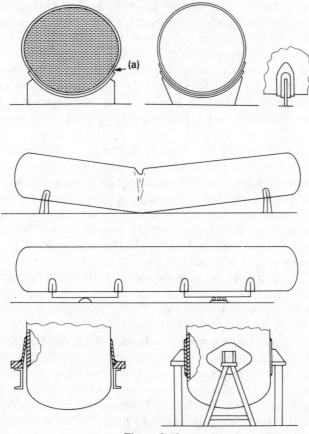

Figure 9.10

9.9 PIPES AND BENDS

In pipework we have to consider stresses due to weight of pipe, not forgetting the contents, lagging, lagging soaked with rain and in special cases also ice. In addition there is thermal expansion. Most pipe runs are laid out with bends, which helps the natural flexibility of the straights and bends together to minimise the end forces and couples due to expansion. It is necessary to calculate the effect of expansion on the stresses. One point that must be remembered is that the intermediate supports must not be allowed to become new fixing-points. They should allow rolling or sliding motion. If there is any vertical expansion, hangers with compensation should be used, preferably of the constant force type using springs or counterweights.

The calculations are the converse of the frame calculations discussed in section 7.9. We impose a net force on the pipe run plus end moments, unknown at first. Then by the moment area method and the energy equation, we find the deflection due to the system using the end conditions of no slope change. The deflection is made equal to the expected expansion due to temperature. In some cases we can guess the points of inflexion, with the help of cardboard models. Then we can reduce the system to a set of cantilevers and add up the strain energy using the standard expression from figure 7.2. The force must run in straight lines through all the points of inflexion; the maximum moment is then easily seen, force times distance from line of force to appropriate corner. The total sum of the strain energies is $= P\Delta/2$. Δ is known, hence P is found, hence bending moments and stresses (see example below).

If we find that stresses are more than trivial we must make a better calculation. This raises a problem of bends. Pipe bends when subjected to a bending moment change their cross-section. If bent outwards they narrow at the flanks, the outer edge (extrados) is pushed out further and the intrados pulls inwards. When flexed inwards to tighten the curve the flanks spread out, the section flattens into an oval tending towards a strawberry shape in extreme cases. This is in-plane bending. In three-dimensional pipe runs we also get torsion and out-of-plane bending.

The ovalising increases the flexibility of the bend compared with an equal length of straight pipe. The ratio of slope change of the bend to that of an equal mean-line length of straight pipe of same diameter and wall thickness is called the flexibility factor, f_f. The ovalising also increases the stress. The ratio of the highest flexing stress in the bend to the highest bending stress in the relevant straight pipe under the same moment is called the stress factor, f_s. Since this stress is at right angles to the normal bending stress which is present anyway, the present work uses a stress factor which directly refers to the highest equivalent tensile stress for yield purposes, that is, the stress intensity as defined earlier (twice the highest shear stress). The hoop stress due to fluid pressure can be added on directly.

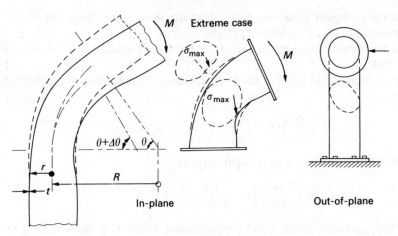

In-plane   Out-of-plane

Figure 9.11

Figure 9.11 shows the situation and notation. We shall only consider in-plane bending. Out-of-plane bending gives similar factors, generally a little lower. The chief references are [57], [58] and [59]. There are theories for pure bends not restrained from ovalising. In practice a pipe bend has either flanges or straight pieces of pipe (tangents) attached to it which reduce the flexibility.

The parameter for flexibility is r^2/Rt (mostly quoted in reciprocal form, Rt/r^2 which is misleading since it is infinite in a straight pipe). Reference [57] contains the greatest range of results but with a few anomalies; the most notable is the effect of flanges immediately at the end of the curved part; the flexibility factor for a 180° bend is given as lower than that for a 90° bend whereas with other end conditions the 180° bend is more flexible than the 90° bend comparing like with like. Reference [58] does not deal with flanged bends; it gives more stress details than reference [57] and uses a more realistic set-up not allowing the third axis freedom allowed in reference [57]. Where comparisons are possible, agreement is satisfactory. The recent reference [59] seems to resolve the anomaly; it agrees with reference [57] on flanged 90° bends and gives self-consistent values for 180° bends. It is particularly valuable in providing data for bends with short straight tangents flanged at the ends of the straights since this is the way standard bends are normally made. That kind of aspect tends to escape the academic investigator.

The data become highly non-linear when Rt/r^2 falls below 0.06. In practical pipes it is almost always well above 0.1. For the region above 0.1 the author has reviewed the available data and finds good agreement with the following.

Take the flexibility factor for a pure bend as $f_f = 1 + 1.8r^2/Rt$, but a long straight tangent of same size and wall thickness as the bend is equivalent to a 9° reduction per end a straight tangent as above, half a diameter long with a flange on its end is equivalent to an 18° reduction per end

a flange immediately on the end of the curve is equivalent to a 36° reduction per end.

For example, a 90° bend of $Rt/r^2 = 0.1$, flanged at both ends, has a flexibility factor

$$1 + \left(\frac{1.8r^2}{Rt}\right)\left(1 - \frac{72}{90}\right) = 1 + 18 \times 0.2 = 4.6$$

A 180° bend with two long straight tangents

$$f_f = 1 + \left(\frac{1.8r^2}{Rt}\right)\left(1 - \frac{18}{180}\right) = 17.2$$

A '135° included angle' bend (conventional notation in some catalogues), actual angle $\theta = 45°$, with two flanged ½ diameter straights, $Rt/r^2 = 0.1$

$$f_f = 1 + 18\left(1 - \frac{36}{45}\right) = 4.6$$

For stresses one should expect a similar relation since ovalising has similar influences on both flexure and stress. The following formula relies mainly on the data in reference [57], giving a stress factor for the highest flexing stress in the transverse plane $f_s = r^2/Rt$ for a pure bend, reduced in the same proportions as above according to the end conditions. For maximum stress intensity, add 1. For example, in the first case above, stress factor $= 1 + 0.2 \times 10 = 3$. The second case gives $1 + 9 = 10$, the third 3.

To summarise these as formulae, if the end corrections above are called c_1, c_2 etc, $c = \frac{1}{2}(c_1 + c_2)$ for any given pipe size.

$$f_f = 1 + \left(\frac{1.8\,r^2}{Rt}\right)\left(1 - \frac{2c}{\theta}\right) \tag{9.15}$$

$$f_s = 1 + \left(\frac{r^2}{Rt}\right)\left(1 - \frac{2c}{\theta}\right) \tag{9.16}$$

c can be any of the above, for example, one long straight and one immediate flange would give $2c = 9° + 36° = 45°$.

It is evident that this rule is too simple to hold when $2c$ gets too close to θ since f cannot be less than 1.

The straight pipe length with which the bend is compared is the whole curved length measured along the centre-line of the pipe, excluding any extra straights, but *not* with the corrections c chopped off.

Example 9.3 (Example of suggested method)

Figure 9.12 shows a pipe fixed at the ends as shown. It is of 200 mm inside diameter, 8 mm wall thickness, of aluminium alloy with $E = 70\,000\,\text{N/mm}^2$, α

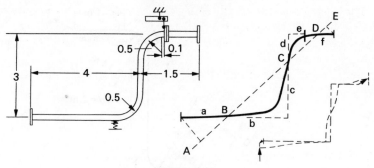

Figure 9.12

$= 24 \times 10^{-6}/°C$. After installation it is cooled to 200 °C below the installed temperature.

First guess: Treat corners as rigid. Estimate line through points of inflexion, ABCDE. Scale off distances to find typical bending moments. Treat a, b, c, etc., as cantilevers with maximum moments M.

a   length 1.7 m, $M = 1.15P$, energy (figure 7.2) $= M^2L/6EI = 0.38P^2/EI$
b   length 2.3 m, $M = 1.5P$
c   length 2   m, $M = 1.5P$, energy for b + c $= 1.61P^2/EI$
d   length 0.9 m, $M = 0.7P$
e   length 1   m, $M = 0.7P$, energy for d + e $= 0.16P^2/EI$
f   length 0.5 m, $M = 0.4P$, energy for f $= 0.01P^2/EI$

Total energy $= 2.16P^2/EI$. This is $= P\Delta/2$ where Δ is the thermal contraction of AE. $AE = 6.26$ m. Therefore

$$\Delta = 6.26 \times 200 \times 24 \times 10^{-6} = 0.031, m \qquad 2.16P^2/EI = 0.031P/2$$

Therefore

$$P = 0.0072EI$$

Highest bending moment $= 1.5P$

$$\text{highest bending stress} = My/I = 1.5 \times 0072EI \times 108/I$$
$$= 82 \times 10^6 \, N/m^2$$

Since stress in pipe bend is appreciably greater than general stress, this is excessive and a fuller analysis is required.

The fuller method proposed here is considered highly suitable when the bends occupy a relatively short part of the total length so that an average line through each bend can be used without great loss of accuracy. Figure 9.13 shows the pipe centre-line, the line for the end force between the terminals and lines parallel to it through the quarter points of the bends. The first step is to find the end moment M_0 in terms of P using the requirement that if the ends

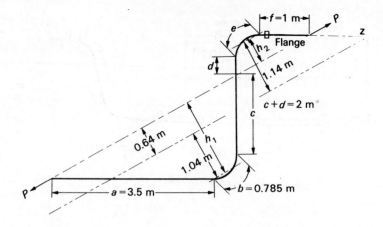

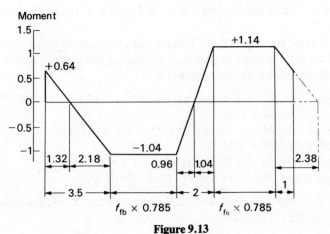

Figure 9.13

are fixed the slope change is zero. The moment area method makes this relatively quick. Then the energy method is used to find the flexural energy in the pipe in terms of P. Finally this is equated to $P\Delta/2$ when Δ is the thermal contraction of the pipe run (end-to-end).

(1) Find the flexibility and stress factors for the bends

$$\frac{r^2}{Rt} = \frac{100^2}{500 \times 8} = 2.5$$

Bend b

$$f_{\text{fb}} = 1 + 1.8 \times 2.5 \times \left(1 - \frac{18}{90}\right) = 4.6 \qquad f_s = 1 + 2.5 \times 0.8 = 3$$

Bend e

$$f_{fe} = 1 + 1.8 \times 2.5 \times \left(1 - \frac{27}{90}\right) = 4.15 \; f_s = 1 + 2.5 \times 0.7 = 2.75$$

(2) Slope change due to P alone, in separate parts.
Lengths a and c have average moment $Ph_1/2$, bend b has moment Ph_1;
Lengths d and f have average moment $Ph_2/2$, bend e has moment Ph_2;

$$\theta \text{ due to } P = P\left[h_1\left(\frac{a}{2} + bf_{fb} + \frac{c}{2}\right) - h_2\left(\frac{d}{2} + ef_{fe} + \frac{f}{2}\right)\right]$$

notice the sign change

(3) Slope change due to M_0 alone.
M_0 is constant throughout, hence

$$\theta \text{ due to } M_0 = M_0(a + bf_{fb} + c + d + ef_{fe} + f)$$

(4) By setting the slope changes equal we obtain M_0, since total slope change $= 0$. Therefore

$$M_0 = 0.64P$$

(5) The strain energy can be added up by standard cases. Taking the lengths from the drawing, each triangle has strain energy $P^2L^3/6EI$ or $M^2L/6EI$ with M as the maximum moment for the triangle concerned. The bent parts are given constant moments of $(1.68 - 0.64)P$ and $(0.5 + 0.64)P$, over a length equal to the length of the bend times its flexibility factor. The problem of the right hand end is overcome by extrapolating to an artificial point Z and subtracting the excess.
The energy sum is

$$[0.64^2 \times 1.32 + 1.04^2(2.18 + 0.96) + 1.14^2(1.04 + 2.38) -$$

$$0.64^4 \times 1.38]\frac{P^2}{6EI} + (1.04^2 \times 0.785 \times 4.6 + 1.14^2 \times$$

$$0.785 \times 4.15)\frac{P^2}{EI}$$

$$= (1.303 + 4.070)\frac{P^2}{EI} = \frac{5.373P^2}{EI}$$

As before, the contraction $= 0.031$ m

$$U = \frac{P\Delta}{2} = \frac{5.373P^2}{EI} \quad P = \frac{0.031EI}{2 \times 5.373} = 2.88 \times 10^{-3}EI$$

The highest stress will be in bend b; its moment is slightly lower but its stress factor is higher.

$$\sigma = 3 \times 1.04P \times \frac{y}{I} = 3 \times 1.04 \times 2.88 \times 10^{-3}EI \times \frac{0.108}{I}$$

Remembering that for this metal, aluminium alloy, $E = 7 \times 10^{10}$ N/m²

$$\sigma = 3 \times 1.04 \times 2.88 \times 7 \times 0.108 \times 10^7 \text{ N/m}^2 = 68 \text{ N/mm}^2$$

(stress factor included)

This is likely to be acceptable for the job, depending on the exact alloy being used.

Comment: for steel, the coefficient of expansion is lower while E is higher. The result would have been similar in stress terms. Can you guess why this example uses aluminium?

9.10 CORRUGATED BELLOWS

When the flexibility of a pipe is insufficient we can use corrugated flexible bellows. These are obtained from specialist makers. There are two kinds of duty, very occasional flexing and regular cyclic duties. Most bellows are designed for 10 000 cycles endurance. In short-life cases larger deflections are permissible; the usual limits for 10 000 cycles life are extension of 10 per cent of the corrugated length with compression of 15 per cent. It is also permissible to use bellows in bending; the bellows should be slightly pre-compressed and the bending restricted so that the shortest side when bent is 15 per cent shorter than the free length but the longest side is not much longer than when it was free. Bellows are not available for very high pressures, since the material must be thin to be flexible. Two, three and even four plies are feasible.

It is important to realise that bellows have relatively little strength in tension. We cannot be sure in an installation that the axial force due to pressure will be supported fully at all times, nor can we be certain of lateral support, as was shown by the Flixborough disaster. If lateral support is lacking, a bellows can buckle like a strut. A length of pipe with a bellows unit each end can also buckle (squirm). References 60 and 61 discuss these forms of failure.

The usual forms of restraint are shown in figure 9.14. The simplest forms a, b are sufficient for many cases. For large expansions a layout such as c is used. The stresses in the hinge-type support may be found as follows.

Find the pressure force, pA, and the maximum possible angle in service. The tensile stress at the hole is easily found. At each arm the force $= pA/2$, the net area is found from the dimensions. It is not often necessary to consider fatigue, hence stress concentration can usually be ignored here. The other place of interest is at the base of the lug. Here we have tension and bending combined. The bending moment is due to the angle; at each lug the bending

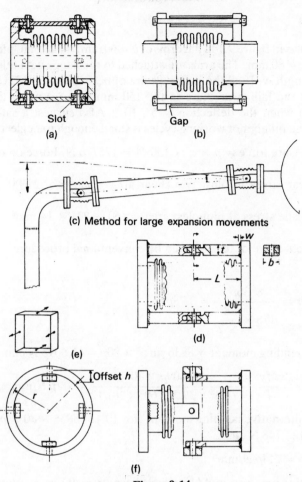

Slot · Gap

(a) · (b)

(c) Method for large expansion movements

(d)

(e)

Offset *h*

(f)

Figure 9.14

moment $= \frac{1}{2}pAL \sin \theta$. However we should consider the general support conditions; in some layouts the *whole* force could come on *one* arm.

The stress depends on the form of the connection. If solid, the bending stress $= 6M/b^2t = 3pAL \sin \theta/b^2t$; the tensile stress $\frac{1}{2}pA/bt$ must be added to this. If the lug is attached by fillet welds of leg length w, the stresses are calculated assuming the throat thickness $w/\sqrt{2}$ to act in tension.

$$\text{Bending stress} = \frac{My}{I} = \frac{1}{2}pAL \sin \theta \times \frac{\frac{1}{2}b'}{(b'^3t' - b^3t)}$$

(see figure 9.14d)

where $b' = b + W\sqrt{2}$, $t' = t + W\sqrt{2}$.

Example 9.4

A hinge unit as in figure 9.14d has arms of $b = 60$ mm, $t = 28$ mm, pin diameter $= 22$ mm, $L = 80$ mm. The arms are attached to the flange with fillet welds of 8 mm leg length w. If all the materials are of $\sigma_y = 220$ N/mm², UTS $= 300$ N/mm² find the failing pressure in the 150 mm pipe (a) when the joint is straight, (b) when the deflection $= 5°$, $10°$. Also suggest a safe working pressure if the number of working cycles is small enough to neglect fatigue.

(1) Pressure force $= p \times \pi \times 150^2/4 = 17671p$ N. Force on one arm $= 8836p$ N.

(2) Shear stress in pins: two shear areas each $\pi \times 22^2/4$; $\tau = 11.6p$ N/mm².

(3) Tensile stress at pinholes: area $= (60 - 22) \times 14$ mm²; $\sigma = 16.6p$ N/mm².

(4) Stress at weld: tension area by conventional procedure $=$ length of weld $\times w/\sqrt{2}$

$$\sigma = \frac{8836p}{178 \times \dfrac{8}{\sqrt{2}}} = 8.78p$$

bending moment $= 8836 \sin 5° \times 80p = 61\ 609p$ N mm

$$\sigma = \frac{My}{I} = \frac{61\ 609p \times 30}{(71.3^3 \times 39.3 - 60^3 \times 28)/12} = 2.7p \text{ N/mm}^2$$

alternative bending moment (for 10°) $= 8836 \times 80 \times p \sin 10° = 122\ 748$ N mm

$$\sigma = 5.4p \text{ N/mm}^2$$

(5) Failing pressure could be when shear stress at pins reaches about $^2/_3$ UTS, or when tensile stress at holes reaches UTS; this will occur when $p = 18$ N/mm²; stress at weld is lower.

(6) Suggested working pressure 4.5 N/mm².

The stresses in a gimbal ring are more difficult to find, particularly the torsional component; a logical way of making a gimbal support would be a square, figure 9.14e. This is four simple beams joined at the corners. A simple cardboard model confirms that such a frame is relatively free from twist. To visualise the action of a ring, a cardboard model is easily made and shows the twisting distortion. The exact form depends somewhat on how and where the pins are fixed. As an estimate we assume that midway between hinges there is a point of inflexion; the 45° arc acts as a curved cantilever of maximum torque $\frac{1}{4}Ph$ where $h = r(1 - 1/\sqrt{2})$ (since we could replace the arc by a cranked arm). The shear stress is then obtainable by equation 2.21. There is still a bending moment of magnitude $\frac{1}{4}Pr/\sqrt{2}$. The stresses from bending are high at just those parts of the cross-section where the torsional stresses are low in a rectangular member. For a rough estimate we may treat them separately; for combination we remember that the shear stress is highest at the centre of the short side. Elsewhere it is fair to take the surface shear stress as inversely proportional to the distance from the centre of the cross-section.

Example 9.5

If a gimbal ring of mean diameter 220 mm, length 60 mm, thickness 28 mm, holes 22 mm diameter, carries an axial force of 80 000 N, estimate the torsional and bending stresses separately. Then find the maximum shear stress at the shorter edge and the highest equivalent tensile stress $2\tau_{max}$. Ignore stress concentration (steady loads). Also estimate the stress in the fillet weld, taking the arm to be 60 mm wide, 20 mm thick, the weld being an 8 mm fillet as in figure 9.14(d).

(1) Find offset: $h = 110 (1 - 0.707) = 32.2$ mm. This is an upper bound assumption since if the joint is arranged as shown the load point will be nearer the inner edge; however some commercial designs have the arm outside the ring, thus increasing the torque.

(2) By equation 2.21

$$\tau = \frac{\frac{1}{4} \times 80\ 000 \times 32.2 \times [3(60 - 22) + 1.8 \times 28]}{(60 - 22)^2 \times 28^2} = 93.5\ \text{N/mm}^2$$

This is slightly high for a shear stress.

(3) Bending moment $= \frac{1}{2} \times 80\ 000 \times \frac{r\sqrt{2}}{4} = 1.5556 \times 10^6\ \text{N mm}$

$$I = \frac{28 \times (60^3 - 22^3)}{12} = 4.8 \times 10^5\ \text{mm}^4$$

$$\sigma = \frac{My}{I} = \frac{1.5556 \times 30}{0.48} = 97\ \text{N/mm}^2$$

Torsional stress at short edge is inversely proportional to distance from centre

$$\tau = \frac{95 \times 14}{30} = 44.3\ \text{N/mm}^2$$

By equation 2.39, $\sigma_e = 131$ N/mm^2 at short edge. But maximum σ_e on long edge $= 2 \times 93.5 = 187$ N/mm^2. Hence somewhere near there, maximum stress slightly higher still.

(4) Stress in fillet weld: take combined bending and tension. Assume pin fixed in arm, free in ring

$$\text{bending moment} = \frac{1}{2}P (10 + 14 + C)$$

where C is clearance between arm and ring which is needed to allow rotation; say 2 mm. Offset = 26 mm. $M = 40\,000 \times 26$ Nmm. Effective fillet = throat length = $8/\sqrt{2} = 5.656$ mm

$$I = \frac{(71.3 \times 31.3^3 - 60 \times 20^3)}{12} = 142\,000 \text{ mm}^4$$

$$\text{Bending stress} = \frac{My}{I} = \frac{40\,000 \times 26 \times 10}{142\,000} = 73 \text{ N/mm}^2$$

$$\text{Shear/tensile stress} = \frac{40\,000}{(2 \times 60 + 2 \times 20)} \times 5.656 = 44 \text{ N/mm}^2$$

in fillet weld

Maximum stress = 117 N/mm²

Rather high for a fillet weld, acceptable in the given arrangement as flexing would reduce bending moment.

If the pipework could have an appreciable deviation as in example 9.4, we would check for stresses due to the angular component of the pressure force.

The importance of sufficient restraint is emphasised by the Flixborough disaster. This was so extensive that it killed all the people who knew details of the plant. From the remains it was clear that a major contributor was a pipe, angled at about 15° between unconfined bellows units, figure 9.15. The pipe itself was restrained by scaffold clamps and short sections of scaffold pipe, which operated successfully for a time. As it happens these bellows were just about strong enough to stand the pressure even if the restraint had given way slowly. It is thought by some that another event disturbed the set-up so that the pipe was released suddenly; this was enough to break the bellows and release large amounts of highly explosive vapour.

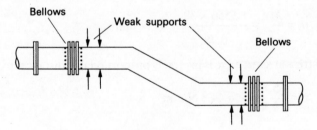

Figure 9.15

10 SHORT CASE STUDIES

10.1 THE COMET Mk 1 [63]

This study shows the importance of residual stress in fatigue and, even more, the importance of ensuring that test procedures truly resemble service conditions. The De Havilland Comet was the first pressurised airliner designed in the United Kingdom. The fuselage was treated as a pressure vessel under repeated loading since it was blown to a pressure of about 0.9 bar when the outside pressure was approximately 0.16 bar. Because there was some uncertainty about SCF at window frames, etc., a test fuselage was built, tested statically to twice the intended working differential P and then fatigue tested to just over P and back to zero, many times. It was kept on test long after the necessary number of cycles had passed.

Meanwhile several aircraft went into service; after a time two exploded in flight. From the recovered pieces it was concluded that fatigue cracks had grown until they reached catastrophic size so that the fuselage burst open although neither craft had suffered as many pressure cycles as the test fuselage.

Question 1. Was there any significant difference between test and working fuselages? Answer: It transpired eventually that the working fuselages had not received a test to double working pressure, only to 1.3 times P. The $2P$ test had washed out the major stress concentrations and put in favourable residual stresses. The $1.3P$ test had much less effect. This was fully confirmed by fatigue-testing fuselages which had only a $1.3P$ pre-test. The work shown in figure 3.14 was subsequent to these events.

Question 2. Why was this effect not known previously? Answer: It was, but in another field; the pressurising of gun barrels (autofrettage) had long been used to lengthen the fatigue life of guns, postponing the onset of cracking.

10.2 TURBINE SHAFT FAILURE

This story illustrates the effect of shrink-fits and the uncertainty of design calculations.

Some low-pressure rotors of a range of 60 MW steam turbines failed in service. The general form is shown in figure 10.1, together with the supposed bending stresses. The main novel feature was in these stresses which had previously been kept down to 23 N/mm². It should be noted that these stresses depend strongly on the setting of the centre bearing which in this type of machine is deliberately below the geometric centre-line to even out the bearing loads.

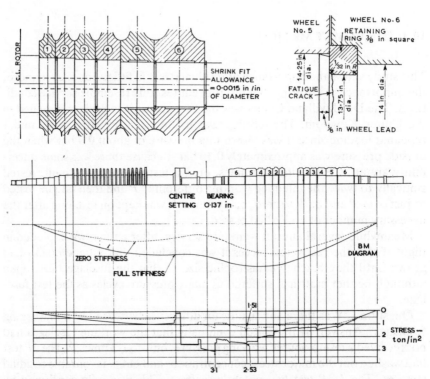

Figure 10.1 Reprinted by permission of the Council of the Institution of Mechanical Engineers from M. B. Coyle and S. J. Watson, Fatigue strength of turbine shafts with shrunk-on discs, *Proc. Instn mech. Engrs*, **178**, 1 (1963–4) 148, 173

The shaft design included some needlessly sharp locating grooves with 0.8 mm corner radii; nevertheless the SCF for the shaft alone was never more than 5.5. The authors of reference 10 make it clear that the effect of shrunk-on bodies was not taken into account. The endurance limit of the material was determined by the usual laboratory tests as ± 370 N/mm². Designing to a nominal stress of 31 N/mm² seemed safe enough, since 5.5 × 31 is only 170.

If we take the added bodies into account, creating in effect a much deeper groove, the SCF comes to about 10½ in bending. In chapter 4 we find that the data do not cover a deep flat-bottomed groove but by interpolation of various data the value of 10½ is obtained. We are however looking for a factor of 12 if we wish to explain the failure. Some of the discrepancy lies in the endurance limit. As seen in figure 3.6, adapted from reference 10, the steam atmosphere and the size bring the endurance limit down appreciably. The actual shafts were about 350 mm diameter; this is much too large for fatigue testing machines. It is generally thought that size effect stops above 2½ to 3 inches diameter although one paper still found some effect up to 6 inches, in coarse-grained material. Allowing for both size and steam effect we still cannot justify the failure. The steady torque would provide a stress of 18 N/mm² without concentration. Adding this on as a mean stress just about brings us to the endurance limit. To explain the *early* failure we must allow for torque fluctuation in the alternator (or include some stress concentration with the steady torque). Data for this are not available in the report.

Comment: the design was changed to semicircular grooves, thus greatly increasing the radius. Other makers tended to use a flat-bottomed groove but with a 6 mm radius.

Question 1: Are we sure of the fillet radius in the grooves? This is not the type of detail that inspectors are unduly interested in provided the radius is *small* enough not to foul the mating part.

Question 2: Should the designers in or around 1960 have been aware of references 25, 26 and several others dating from 1932 to 1935?

Question 2a: Why do the usual stress calculation textbooks fail to emphasise this effect?

10.3 DIESEL ENGINE CONNECTING ROD [1]

Figure 1.1. showed a connecting rod splitting on a line normal to the main stresses. A connecting rod has compressive stress due to the force of the burning gases; in a diesel engine this is the highest stress. Other stresses are tensile due to decelerating the piston on the upstroke and accelerating it downwards. These are the greatest in high-speed four-stroke Otto engines at exhaust top dead centre. Then there are bending stresses as in section 7.4. The rod design in question came from a large low-speed diesel engine so that bending stresses would be small. At first, forging defects were suspected. Then it was thought that the top of the rod might have been fracturing due to the tensile force, but calculations showed this force to be too small to explain the failure. Finally some rods were found to be cracked at the stem but without cracks at the rod top. The explanation was found by considering that the downward force was applied through the oil film between rod and gudgeon pin. This makes it similar to a pressure vessel with a radial pressure. The rods all had a central oilway which gave a stress concentration factor of

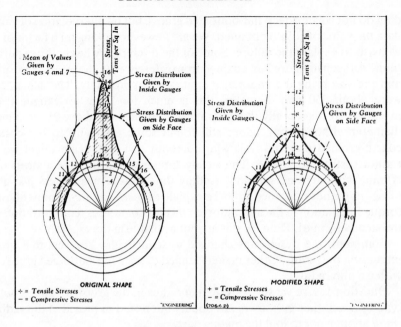

Figure 10.2

2.5 to 3. This gave understandable stresses by calculation, duly confirmed by measurement (figure 10.2). Then by reinforcing the opening as in pressure vessel practice further trouble was avoided.

10.4 ROTOR PIN STRESSES

A fan rotor as in example 6.5 is shown in more detail in figure 10.3. It is made of material of density 4500 kg/m³ and rotates at 1432 rpm. It is required to find the stress in the pins (modified from an actual case where the pins fractured).

The speed is 1432 rpm, the material's density = 4500 kg/m³, $E = 1.1 \times 10^{11}$ N/m², $\nu = 0.33$. The plate thicknesses are 15, 10 and 5 mm, the blades are 5 mm thick. The first plate is 400 mm diameter, the eye is 600 mm and the outside 1500 mm. The pins are 12.7 mm diameter, 120 mm long, placed at 580 mm radius. The shroud slope is 1 in 4.

(1) Transverse load on pins

$$\frac{\text{mass}}{\text{unit length}} = \frac{\pi \times 12.7^2 \times 4.5 \times 10^{-6}}{4} = 5.7 \times 10^{-4} \text{ kg/mm}$$

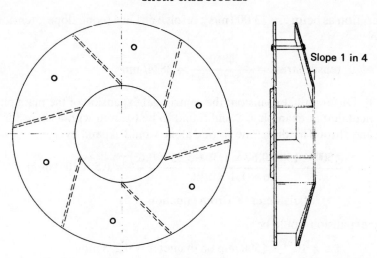

Figure 10.3

$$\text{Acceleration} = \omega^2 r = \left(\frac{2\pi \times 1432}{60}\right)^2 \times 0.58 = 13\,000 \text{ m/s}^2$$

Loading per unit length = 7.4 N/mm

Bending moment depends on support condition: the pin is fixed to the plates but the plates are relatively thin; the built-in assumption is an upper bound giving $M = wL^2/12$ (section 7.5)

$$\text{Bending stress} = \frac{My}{I} = \frac{32M}{\pi d^3} = 44 \text{ N/mm}^2$$

at fixing, SCF not included yet

(2) Tension in pins: it is argued that the centrifugal acceleration of the sloping shroud plate is supported by the hoop stress in the direction of the cone only; if there were no pins and no blades the shroud would flatten out significantly. The pin is assumed to support a certain area of shroud estimated at 220×300 mm

Pin tension = mass of given shroud area $\times \omega^2 r \times \tan\theta$
(Note that $\omega = 150$ rad/s)

mass = $4.5 \times 10^{-6} \times 220 \times 300 \times 5 = 1.485$ kg acceleration as before = 13 000 m/s²; resolve at 1 in 4 (cone slope); tension = 4850 N

$$\text{tensile stress} = \frac{4850}{\pi \times 12.7^2/4} = 38 \text{ N/mm}^2$$

(3) Differential expansion: the centrifugal expansion of the main plates was calculated in example 6.5 and found to be 0.2 mm with a share for the blade and shroud loading. The shroud alone would expand by

$$\frac{4500 \times 150^2 \, (0.67 \times 1.5^2 + 3.33 \times 0.6^2)}{16 \times 1.1 \times 10^{11}} = 1.5 \times 10^{-4} \text{ m/m}$$

of diameter   (from equation A.5)

the tip expansion would be

$$1.5 \times 10^{-4} \times 1500 \text{ mm on diameter} = 0.225 \text{ mm}$$

With a share of the blade mass to be carried this would go up to say 0.3 mm, but by the same argument as for the main disc we should share out the blades between shroud and disc when looking for the differential. This leaves a discrepancy of approximately 0.1 mm on diameter, or 0.05 mm radially. The effect of the blades is to bend, thus introducing a relative rotation between disc and shroud, with the blades at 45° this is also about 0.05 mm. Thus the pins are displaced by 0.71 mm transversely. The stress introduced by this is estimated by considering half a pin as a cantilever with a tip deflection of ½ × 0.71 mm

From figure 7.4

$$0.0355 = \frac{P \times 60^3}{3 \times 1.1 \times 10^5 \, (\pi \times 12.7^4/64)}$$

and

$$\text{bending stress} = \frac{P \times 60}{(\pi \times 12.7^3/32)} = 21 \text{ N/mm}^2$$

(4) Conclusion: these stresses concur at the fixing, hence

total tensile stress = 21 + 38 + 44 = 103 N/mm²

This is not enough to explain fracture in a plain bar but in the presence of a fillet weld and vibration adding to the tensile and flexural components, failure becomes credible.

Comment: the pin method is not uncommon in mild steel fans; when high-tensile, notch-sensitive materials are employed or when corrosive conditions are present a modified pin design may be used, separating the weld from the geometric stress-raising corner (figure 10.4).

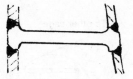

Figure 10.4

10.5 MINE HOIST BRAKE FAILURE

This study is based on an actual case but since the author wishes to stress certain points not fully detailed in the report it is not necessarily true to life quantitatively and is thus to some extent fictitious. The mine hoist considered here had one retarding system incorporated in the electric drive (regenerative braking), backed up by a fail-safe friction brake on the cable drum. These brakes are always fail-safe, being applied by a dead-weight or in this case by a nest of many springs, through a lever system. The brake is held off by a pneumatic system so that if power fails the brake comes on automatically independently of external action. Some readers may be surprised to note that no use is made of the four stout guide ropes for emergency brakes. At the time of writing a hoist with the full safety features normal in passenger lifts is being installed in a coal-mine, believed to be the first in Britain and possibly the first in any mine.

The layout is shown in figure 10.5a and b. The system is roughly balanced when at mid-depth, though naturally the loads in the two cages or cars are not always equal. The weight of the paid-out length of rope accelerates the system from the mid-point towards the extremes. A normal trip starts with applying enough current to overcome the imbalance, releasing the brake, accelerating, then winding at steady speed governed by a mechanism or by the winch-man, then electric retard and finally letting the brake come on. On the occasion under discussion, the brake rod fractured during brake application, the cages accelerated out of control. The rising cage was arrested in the headgear by the safety chains, the other crashed into the end stops with many fatalities. The governing gear was found to be damaged afterwards which may have been consequential since the whipping of the overrunning rope caused additional damage.

The rod R was found to have several cracks, originating at the thread root which is the obvious weak spot, one of which had progressed over half-way as a fatigue crack before finally parting over the remaining cross-section. All the cracks were on one side, indicating that bending was the likely cause. The pivot block bearings were found seized though it was not clear whether they were in the brake-on, brake-off or medium position. Tests with a fresh rod showed that in addition to the tensile stress of some 60 N/mm^2 on the thread root area an additional bending stress of $\pm$ about 90 N/mm^2 occurred in the

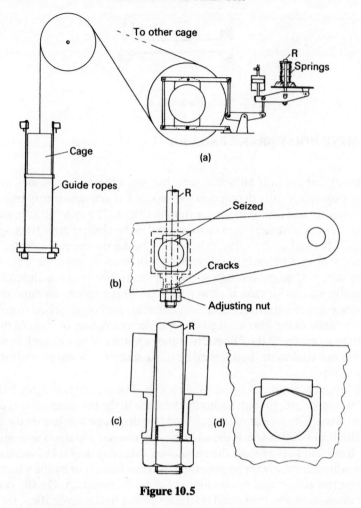

Figure 10.5

rod during brake application, in the direction compatible with the crack positions. It is not important in which position the pivots had originally seized, presumably in the resting state with the brake on, during which the high contact pressure squeezed out the lubricant. The actual rod angle would change as the linkage was adjusted to compensate for brake lining wear.

The bending action is illustrated in figure 10.5c, in exaggerated form. The reader should by now be able to confirm that if the load were really taken entirely on the point shown, the bending stress would be between 5 and 6 times the nominal tensile stress, which would take it well beyond yield for the material (0.4 per cent carbon steel) if the SCF is taken into account. We conclude that either the material was repeatedly taken up to yield point locally (see section 3.3) or that partial support was given in the pivot block.

Design lessons: what seemed to be a very sound design is now seen to have failed. It need not have failed if any one of the following had been incorporated.

(1) A pressure-greasing facility for the bearings;

(2) A rocking pad instead of the bearing provided, (figure 10.5d);

(3) Much less clearance in the pivot block – this would have concentrated the flexing in the plain part of the rod, away from the stress concentration due to the screw thread. The rod was 2.4 metres long, 50 mm diameter, and could have comfortably stood angle changes of $\pm 5°$ at one end indefinitely, the other end being relatively free thanks to the spring nest;

(4) Possibly a spherical seating for the adjusting nut; this would have presumably become unseized every time the brake clearance was adjusted, provided that the seating was pinned so as to be unable to turn with the nut.

APPENDIX
(Mainly of fourth year or Master's degree standard)

A.1 TIME AND SPEED EFFECT IN FATIGUE TESTING

This aspect is crucial to the validity of accelerated life testing. In the early days the importance of machine inertia effects was discovered when some slack developed in a machine; the life in cycles at supposedly the same loading was much shorter at higher speeds. This problem was speedily disposed of; in the types of machine and specimen commonly used it was confirmed that high speed did not give seriously different results from moderate speed.

In most fatigue testing machines the force is monitored by a dynamometer bar which is a dummy specimen of larger area than the test-piece, fitted with strain gauges. As discussed in reference 64 this does not give the force in the specimen since the force to accelerate the (often quite massive) specimen grip must be added to or subtracted from the measured force before we can know the specimen load. Most machines fit the dummy at the fixed end so that the specimen load is greater than indicated by the gauges. This error is likely to be significant in very small specimens, at very high frequencies as in vibratory machines, or in torsion work where the polar moment of the grips may be large. Since inertia errors are not widely understood and details of speed, grip mass, etc., are rarely given, many results may be erroneous to some extent.

In addition to inertia errors, there is another speed effect which was not important in the early days; this is the heating effect. In very small specimens the heat energy is readily conducted away; when we come to larger specimens, around 25 mm diameter, the authors of reference 65 reported significantly higher temperatures at 2900 rpm than at 1900, enough to affect the endurance limit values.

The genuine time effects are most important in low-cycle, high-stress work. Wöhler [8] compared results in rotating bending at 72 rpm with those under an interrupted rotation with a short stop every 90°, effectively reducing the speed to 18 rpm, still in fully reversed loading, and found a substantial reduction in life. Musuva and Radon [9] have recently extended the frequency range and confirmed the reduction of life at low speeds. This is a major handicap, casting serious doubts on many designs based on accelerated tests.

A.2 SIZE EFFECT IN FATIGUE

As noted earlier, in rotating bending small specimens regularly show higher endurance limits than large ones, due to stress gradient. This effect is misleading both in design and in failure investigations. To obtain results unaffected by size we should use specimens of a size reported variously as 60 mm up to 150 mm; the best way is to use push–pull testing which of course requires great precautions against stresses due to misalignment during the compressive part of the cycle. Even more important is the need to avoid inadvertent prior loading while tightening the specimen in the machine, or when switching on and setting up the load cycle.

The basic arguments and explanation of size effect are 60 years old. If we go back to Griffith's view that the region of interest is a few grain diameters below the surface we can explain most of the observed effects. These are particularly important in the neighbourhood of notches. Frost, [11] working with mild steel and later also with aluminium alloys in rotating bending showed that grooves with a tip radius of about 3 grain diameters gave much the same strength reduction factor (effective SCF) as those with sharper tips, down to 0.1 grain diameter. The specimens acquired non-propagating cracks reaching a depth of about 3 grain diameters and then continued to survive with no further crack growth.

The length of the cracks as seen at the surface varied as the cube of the nominal stress, amounting to some 10 to 15 grain diameters. It was hoped that crack length might be a guide to the crack depth. Frost's pieces were of above 500 grains diameter, with grooves up to 120 grains in depth. In large work the width–depth ratio can be much greater; it is not clear whether this is a matter of stress gradient per unit grain size or whether it is a feature of the particular structure. In pressure vessels we would like to give the cracks (if any) strict orders to grow inward, penetrating deeply, to give a warning leak. If a crack grows mostly in width, it may reach catastrophic dimensions, weakening the structure and giving instant, explosive failure without warning. At the time of writing it does not seem possible to guarantee that a vessel will leak before bursting.

Peterson [24] does not give the grain size concerned by shows the size effect in terms of engineering components such as shouldered shafts, rods with small holes, etc. The general conclusion is that the SCF values obtained from experiments must be viewed with the greatest suspicion unless the size and grain size are declared; theoretical and photoelastic data as in chapter 4 are generally on the safe side, provided that fretting is absent.

A.3 LOADING SEQUENCE AND STRESS RATIO IN FATIGUE

(1) As already mentioned, the process of running-in a specimen at below the endurance limit (understressing) or at progressively increasing loads (coaxing) or even running slightly above the endurance limit for a short period (overstressing) have all been shown to raise the eventual endurance limit substantially (see Forrest [13]);

(2) A recent sequence, repeating this type of work but *in fully reversed torsion*, showed substantially longer survival sums when the tests were run in the order low stress before high stress, but with sums well below par when high stresses came first. [66] Other recent efforts have rediscovered the overstressing effect, with longer life if the first part of the test sequence consists of a small number of high-stress cycles. Fully reversed torsion is rare in practice but has advantages in interpretation, thanks to the simple shear stress pattern;

(3) Miner [15] in his work on cumulative damage included extensive variations of loading cycles. Being interested in aircraft structures he used stress ratios of -0.2, $+0.2$ and $+0.5$ in various mixtures, rather than fully reversed cycles. Under these circumstances the conclusions are *completely opposed* to the torsional ones. Many tests gave survival sums $\Sigma n/N$ close to 1. In every case except one, using high loads first gave survival sums of 1 or substantially above. The exception was a test in which the high loading was maintained for $\frac{3}{4}$ of the expected life. Apart from this exception, the low-before-high tests gave significantly shorter lives with survival sums down to 0.6.

These opposing observations become comprehensible if we consider (a) work-hardening and (b) the effect of a prior load as in figure 3.9 extended down to microscopic high-stress points in the material. The effect of overloads in delaying subsequent cracking has been shown in many papers.

(a) Benignant work-hardening would occur in running-in at loads as high as possible but just below the level of producing damage in the form of incipient cracks. This explains the longer life obtained from low-before-high stress tests provided that the low stress is not too low but is close to the endurance limit; just below for many cycles or somewhat above for shorter periods;

(b) Residual stresses from the first loading (figure 3.3) would be beneficial if the later loading is mainly one-way in the same direction as the first load but if the loading is fully reversed this benefit is nullified. Thus fully reversed loads well above the endurance limit would be likely to cause damage of such size that it would continue to grow under lower loads later on.

The evidence we have of non-propagating cracks is of great practical importance; it means that cracked parts do not necessarily need scrapping. In cases where the stress is regular and predictable, cracks can be tolerated. In duties subject to occasional high loads, for example, aircraft, trains, ships, mechanical presses, etc., the situation is more difficult since it may happen that a casual high load opens a crack up so much that the new crack is above threshold size and will propagate at moderate loads.

(4) The effect of stress ratio with a constant form of spectrum is less definitely known; a small hint may be gathered from figure A.1, adapted from reference 67. The life curves taken from wheel rolling tests under varying

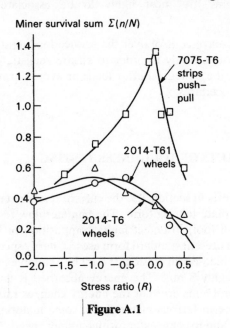

Figure A.1

loads are consistently below the survival sums noted by Miner. There may be a slight environment factor present since Miner used Alclad 24S–T, a Duralumin-like alloy with pure aluminium cladding. The upper curve seems anomalous yet the consistency is high. The author ventures to guess that the hydraulic machine used to test the strip specimens for this curve may have had relatively sluggish changeovers when the lowest load condition was near to zero, thus providing longer rest periods with an increased chance of crack-healing.

From the literature we note several opposing phenomena which will vary in importance depending on time, material, temperature, stress level and whether we have tensile or shear cracks. These include

   (a) time for crack growth;
   (b) time for stress relief (time-dependent yielding, cold or hot creep);
   (c) time for crack healing;
   (d) varying time for a, b and c due to loading wave-form;
   (e) When concerned with low frequencies it is necessary to consider the

possibility of structural changes. These have been investigated to some extent in terms of hold-time. Generally, long hold-times under tension reduce fatigue life; yet in some materials compressive hold-time is harmful while in others tensile hold-time is beneficial. Obviously several factors can enter into this aspect: phase transformations may be encouraged by tension or compression, oxidation may be present, creep may affect crack sizes. The hold-time problems are most likely to be associated with elevated temperatures.

Since we rarely have control over the service loading it is valuable to be aware of these points firstly in order to ensure realistic test programmes, secondly to prescribe beneficial prior loads or avoid harmful prior loads in commissioning procedures.

A.4 SOME ASPECTS OF TOUGHNESS TESTING

As mentioned briefly in section 3.5, the current tests are either the hammer-blow type or the plate-tearing form. The hammerblow test was originally a production control tool, to detect faulty composition or heat-treatment in steel-making. The present standard form uses a Charpy specimen, figure A.2, or if necessary a scaled-down version. It is hit by a suitable hammer pendulum at a speed of roughly 3 m/s. The energy absorbed is the measure of the toughness. The problems are that the energy changes rather abruptly with falling temperature in ferrous materials; in some materials it also changes abruptly with size and to some extent with hammer speed.

Before discussing these items we digress briefly into the action of hammer-blows. In a hammer and nail situation the blow starts a stress wave along the nail, travelling with the rod speed $C = \sqrt{(E/\rho)}$. This wave is reflected (in full

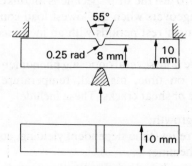

Figure A.2

if the nail does not move forward) back to the hammer, then back along the nail with reduced amplitude, etc. If we hit a block we get a spherical wave, moving slightly faster, speed $\sqrt{[E/(1 - 2v^2)\rho]}$. Occasionally we may get a two-dimensional (cylindrical) wave of speed $\sqrt{[E/(1 - v^2)\rho]}$. In the Charpy specimen the main effect is likely to be a transverse *shear* wave, of speed $\sqrt{(G/\rho)}$. The stress in the first wave front is $\rho u C$ where u is the hammer speed. In the Charpy case it is presumably of magnitude $u\sqrt{(\rho G)}$; note that wave speeds are 3 to 5 km/s.

Large-scale brittle failures are often associated with defects coupled with temperatures a few degrees lower than usual; in the laboratory it tends to take 30 °C to show up a major drop in impact properties. But if a small crack is created first, (see figure A.3, from reference 68 but re-plotted to a common scale) the transition becomes sudden.

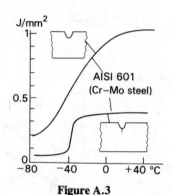

Figure A.3

The safety assessment of pressure vessels against brittle fracture is based on the progressive nature of the fall-off in impact-absorbing capacity with temperature. The lower curve in figure A.3 casts some doubt on this. It is generally accepted that the critical temperatures for brittleness are size-dependent, so it is possible that in thicker work the transition may be equally sudden but at higher temperature.

The effects of size, shape and hammer speed were investigated in a series of tests sponsored specifically in order to establish standards for impact testing. One of the major items in this programme is reference 69. One series investigated width alone, another covered a series of geometrically similar forms while maintaining a common notch root radius and ensuring identical material by the ingenious idea of using a broken half from the largest specimen to make the next, etc. Hammer speed was investigated in a further sequence. All the reported tests were averaged from at least three to four consistent results. The hammer speed made little difference to carbon steel specimens but mild steel was found to be much more brittle at hammer speeds over 8 m/s.

As noted above, the stress in the first wave front $= \rho u C$; with $u = 3$ m/s this is 70 N/mm^2, whereas with $u = 80$ m/s it is 190 N/mm^2. Thus the lower-speed blow would not cause fracture until several stress-reflections later while the fast blow may cause a break at once or at the first wave-return. It is thought that the carbon steel behaves similarly at both speeds because it was more brittle anyway, the finer grain size in relation to notch radius being the governing factor.

Figure A.4 shows the main data from reference 69 replotted in terms of the thickness t and energy per unit area. Fundamentally it should be energy per unit volume in the fracture zone. This volume is not readily assessable; it consists of the cross-section multiplied by 1 or 2 times grain size (which is not usually stated although it would be easy enough to find from the fracture appearance), plus an extra volume of plastic zone related to notch radius.

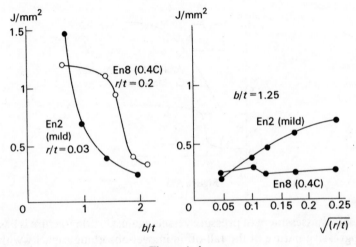

Figure A.4

The plot in terms of b/t ratio is significant since this determines the extent of triaxial tension. The stress due to the primary bending is tensile along the specimen. The curvature of the lines of force gives a transverse tension too, as if trying to delaminate the specimen. Both these cause a Poisson contraction in the third direction. The wider the bar, the more is this local contraction hindered by the adjacent material which is under lower stress. This resistance to contraction induces tension in the third direction which, as we have noted before, encourages fracture.

The graph of energy against $\sqrt{(r/t)}$ seems significant since it brings together both the V notch and the older keyhole notch. For bending loads it may perhaps be more relevant than the radius/notch depth ratio.

To summarise these arguments, both r and t should be such that r/t is not too different from the real structure, r should be many times greater than grain size and b should be many times greater than t to give a good proportion of plane-strained material.

The main forms of fracture (non-impact) specimens are as in figure 3.10. The present object is to debate to what extent these misrepresent the real structure. One aspect is the energy. Depending on notch depth, the local strain energy can well be of the same order as the energy in the rest of the specimen plus some energy in the relatively stiff testing machine. In a long structure the adjacent energy may be much larger. The rate of crack advance could well be augmented by this energy, transported by stress waves, so that real structures fail more readily than predicted. The existence of stress waves in fracture testing has been confirmed experimentally. [74]

The other aspect is stress distribution. In a member with a through-crack the material next to the crack in the load direction is to some extent unloaded. The load goes mainly from the unbroken part to the rigid clamps. The shorter the shear faces SF, figure A.5, the greater is the unloading effect. In the centre-cracked version this effect is still present though somewhat reduced since the Poisson contraction is prevented, figure A.5a to c.

In pressure vessels we are interested in a relatively wide part-through crack. This modifies the unloading because the local extension is much less than with a through-crack of similar area. Also the nature of fluid pressure loading

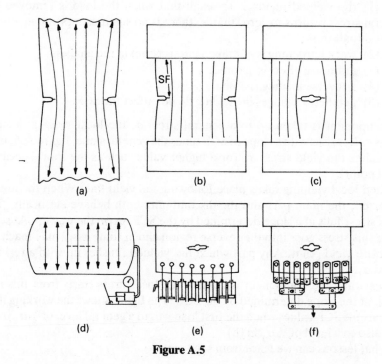

Figure A.5

ensures that load is distributed evenly along the cylinder or sphere, see figure A.5d. To represent this with a through-cracked plate the loading should be by a series of independent cylinders or by equalising harness as used in aircraft structure testing, e and f. In addition we should remember that in a pipe the fluid can enter the crack, causing an additional opening force and moment, the hydro-wedge. There may also be significant chemical activities in the crack. It has recently been shown that distilled water speeds up crack growths substantially. Even if hydro-wedge and part-through effects are absent, plate and cylinder data differ. [20]

The evidence for a size effect as such, whether or not due to energy of adjacent material feeding into the crack, is limited but consistent; for example reference 70 compares tests on 1 m vessels with some of 0.15 m diameter or thereabouts. High stress fatigue life is consistently shorter in the larger vessels.

A.5 RESIDUAL STRESS DUE TO LOCAL YIELDING

In this section we look at the residual stress from local yielding quantitatively. The following simplifying assumptions are made:

(1) the yielded region is so small that when the load is removed the general stress returns to zero, that is, there is no significant propping due to the residual stress;

(2) work hardening is not taken into account in the analysis;

(3) grain size effect is ignored;

(4) a reverse loading is absent;

(5) geometry changes in yielding do not affect the SCF.

A component is taken up to a general stress σ, the local stress at a small stress-raiser rises with a slope of SCF times the general slope, figure A.6, until it reaches the yield stress σ_y (or a higher value such as the short-specimen yield point).

Then local yielding takes place following the yield line. When the load is removed, the main part and the discontinuity both behave elastically. The local stress falls at a slope determined by the SCF, as on the way up. At some stage this stress goes into the reverse region and in some cases may reach the opposite yield point. Any part which has yielded finishes up with a residual stress σ_r.

Next time the component is loaded, the local stress starts from this new level, at the slope determined by the SCF. The figure shows the working lines for various SCF values when the first load was to a general level of $\frac{3}{4}\sigma_y$ in (a) and also to a level of $\frac{1}{2}\sigma_y$ in (b).

What lessons can we learn from such plots?

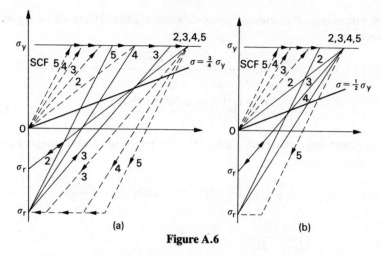

Figure A.6

(1) A prior load to $\frac{3}{4}\sigma_y$ has much more effect in lowering actual peak stresses during subsequent service than a load to $\frac{1}{2}\sigma_y$. If this prior load was $1\frac{1}{2}$ times the working load it has tended to wash out stress raisers up to SCF = 4 by bringing the peak values to well below σ_y. SCFs over 4 will, during cyclic loads, continue to yield at every cycle until they either work-harden sufficiently or fail. The traditional pressure test for vessels to $1\frac{1}{2}$ times working stress has far more effect than the modern one to $1\frac{1}{4}$ times working stress, for stress-raisers up to SCF = 4. For higher SCFs it becomes important to keep the working stresses correspondingly lower.

(2) The second lesson comes from section 10.1; it is the same lesson as (1) but thought of in terms of avoiding false, dangerous experimental results. If a fatigue specimen is unknowingly given a prior load it may give totally misleading, high results. It can be quite difficult to ensure that this does not happen when securing a small specimen in a heavy machine or when starting up a complex control system. The specimen will not show any conspicuous signs of such a prior load.

(3) The high working stresses permitted in bolted joints are explicable on the basis that a first loading induces favourable residual stresses at the roots, then the surfaces settle slightly so that the later loads are appreciably smaller than the initial tightening load, hence even with fluctuations the prior load is not exceeded.

A.6 DISCS AND THICK CYLINDERS

(a) Summary of equations:
Constant thickness disc, angular velocity ω, internal radial stress σ_{r1}, external

radial stress σ_{r2}. All stresses positive when tensile. Hoop stress σ_h at any radius r

$$\sigma_h = \frac{\rho\omega^2}{8} \left[(3+\nu) \frac{r_1^2 + r_2^2 + r_1^2 r_2^2/r^2}{r^2} - (1+3\nu)r^2 \right] -$$

$$\frac{[\sigma_{r1} r_1^2 (r^2 + r_2^2) - \sigma_{r2} r_2^2 (r_1^2 + r^2)]}{r^2 (r_2^2 - r_1^2)} \tag{A.1}$$

At the inner and outer radii 1 and 2 the hoop stresses and diameter changes are

$$\sigma_{h1} = \frac{\rho\omega^2}{16} \left[(3+\nu)D_2^2 + (1-\nu)D_1^2 \right] - \frac{\sigma_{r1}(D_2^2 + D_1^2)}{(D_2^2 - D_1^2)} +$$

$$\frac{2\sigma_{r2}D_2^2}{(D_2^2 - D_1^2)} \tag{A.2}$$

$$\sigma_{h2} = \frac{\rho\omega^2}{16} \left[(1-\nu)D_2^2 + (3+\nu)D_1^2 \right] - \frac{2\sigma_{r1}D_1^2}{(D_2^2 - D_1^2)} +$$

$$\frac{\sigma_{r2}(D_2^2 + D_1^2)}{(D_2^2 - D_1^2)} \tag{A.3}$$

$$\frac{\Delta_1}{D_1} = \frac{\rho\omega^2}{16E} \left[(3+\nu)D_2^2 + (1-\nu)D_1^2 \right] - \frac{\sigma_{r1}}{E} \left(\frac{D_2^2 + D_1^2}{D_2^2 - D_1^2} + \nu \right) +$$

$$\frac{\sigma_{r2}}{E} \left(\frac{2D_2^2}{D_2^2 - D_1^2} \right) \tag{A.4}$$

$$\frac{\Delta_2}{D_2} = \frac{\rho\omega^2}{16E} \left[(1-\nu)D_2^2 + (3+\nu)D_1^2 \right] - \frac{\sigma_{r1}}{E} \frac{2D_1^2}{D_2^2 - D_1^2} +$$

$$\frac{\sigma_{r2}}{E} \left(\frac{D_2^2 + D_1^2}{D_2^2 - D_1^2} - \nu \right) \tag{A.5}$$

(b) The case of a solid disc (no centre hole), under centrifugal action:
The special aspect of this is that the equation for a disc *with* a hole has zero *radial* stress at the centre. When the centre is solid it is no longer possible to have a hoop stress without a radial stress since any tangential expansion automatically implies a radial strain of the same magnitude. Thus the true equation must include the effect of σ_{r1}. This halves the stress as compared with a disc that has a hole, however small. The reader can check this by substituting $\sigma_{r1} = \sigma_{h1}$ in equation A.2 above, σ_{r2} being zero. The working equations were given in section 6.4;

(c) Partial yielding:
A thick cylinder or constant thickness disc can be safely used in the partly yielded state or it can be put into this state temporarily to produce desirable

residual stresses. The situation is best understood graphically. Figure A.7 shows the elastic stresses for a rotating disc and an internally pressurised cylinder, also the partly-yielded stress distribution and the resulting residual stress after load removal. To calculate the speed or pressure needed to produce a given amount of yielded zone we use the equal area concept. The shaded area gives the required force. In the centrifugal case, the force on half a disc, using unit thickness, is $\int_0^\pi \int_{r1}^{r2} \rho \, \omega^2 \, r^2 \sin \theta \, dr \, d\theta$. In the other case force = pD_1 which is easily proved by resolving

$$\text{Force} = \int_0^\pi pr_1 \sin \theta \, d\theta = 2pr_1 = pD_1$$

The dashed curve merely indicates the character of the stress distribution at some lower speed or force, below yield point. For an analysis, the simplest way is to make an arbitrary choice of the radius at which yield is to start. The shaded area is found by an integration plus a rectangle. The residual stress is most readily estimated graphically: the height h is the missing spike. The final state has the lower curve placed such that the areas a and b are equal, hence the lower curve is also required. A graphical method would give a quick approximate answer.

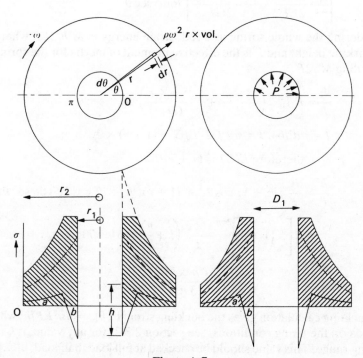

Figure A.7

A.7 BUCKLING OF COMPRESSION SPRINGS, ALSO SURGE FREQUENCY DERIVATION

A.7.1. Buckling Force

To find the buckling force for a spring it is necessary to find the effective I value. In bending a spring about the x axis, figure A.8, the local moment caused by a moment M is $M \cos \phi$; the torque is $M \sin \phi$.

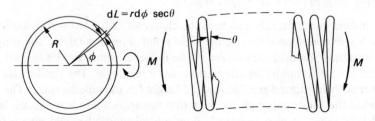

Figure A.8

The strain energy in an element dL is

$$\left[\frac{(M \cos \phi)^2}{2EI} + \frac{(M \sin \phi)^2}{2GJ} \right] R \, d\phi \sec \theta$$

Considering the whole spring as a rod, strain energy $= M^2 h/2EI'$ where h is the working height and I' is the effective moment of inertia for the spring. Cancelling $M^2/2E$

$$\frac{h}{I'} = R \int_0^{2\pi n} \left(\frac{\cos^2 \phi}{I} + \frac{\sin^2 \phi \times E}{GJ} \right) d\phi \times \sec \theta$$

$$I = \pi d^4/64, \; J = \pi d^4/32, \; E/G = (1 + v) \times 2,$$

$$\text{therefore } E/GJ = (1 + v)/I$$

$$\frac{h}{I'} = \frac{R}{I} \int \tfrac{1}{2} + \tfrac{1}{2} \cos 2\phi + (1 + v)(\tfrac{1}{2} - \tfrac{1}{2} \cos 2\phi) \, d\theta \sec \theta$$

$$= R \left[\phi \left(1 + \frac{v}{2} \right) \right]_0^{2\pi n} \frac{\sec \theta}{I} \left(1 + \frac{v}{2} \right) \times 64 \approx 73$$

$$I' = \frac{d^4 h \cos \theta}{73 \, nD}$$

The buckling calculation gives the buckling strength $P_{ce} = n\pi^2 EI'/h^2$ where n depends on the fixing conditions, see section 2.3.3, for any value of h in the working range. This value should be checked at full load, half load, etc., since P and h are different at each condition. Working P should be $\ll P_{ce}$.

A.7.2 Surge Frequency

Surge frequency: by analogy with a rod, where $C^2 = E/\rho$, we require the speed C_s in the spring. $E = PL/A\Delta$, $\rho = m/LA$, therefore

$$C_s^2 = \frac{PL^2}{m\Delta}$$

We take $\sec \theta \approx 1$ in this section. In a helical spring, $m = \rho N\pi D\pi d^2/4$; $\Delta = 8PND^3/Gd^4$
therefore

$$C_s^2 = \frac{4GL^2d^4}{8\,\rho N^2\pi^2 D^4 d^2}$$

$$C_s = 0.22\,\frac{Ld}{ND}\left(\frac{G}{\rho}\right)^{\frac{1}{2}}$$

L is the working length. But $(G/\rho)^{\frac{1}{2}}$ is the speed of shear waves in the material, ≈ 3200 m/s or $126\,000$ in/s for steel. Thus the speed of longitudinal waves in a steel spring $= 700\,d/ND^2$ m/s, if d, D are in metres. The frequency is obtained by the number of times the wave passes up and down again, $f = C/2L$. Thus $f = 350d/ND^2$ Hz if all dimensions are in metres, $350\,000$ for mm units, $14\,000$ for inch units. This is a long-recognised figure, [71] though not as well-known as it should be.

A.7.3 Helix Angle Correction

(a) Rotation can take place freely:
In this case the moment on the wire $= \frac{1}{2}PD$ everywhere. However the wire is in bending as well as torsion. If θ were $90°$ the wire would be in pure bending. Hence the energy is in two parts. The effective length of an element dL is taken as $dL\cos\theta$ for torsion, $dL\sin\theta$ for bending

$$U = (\tfrac{1}{2}PD)^2\left(\frac{L\cos\theta}{2GJ} + \frac{L\sin\theta}{2EI}\right)$$

Equation A.23 gives $E = 2(1 + v)G$. Since $L = \pi nD\sec\theta$

$$\frac{dh}{dP} = \frac{8nD^3\left(1 + \dfrac{\tan\theta}{1+v}\right)}{Gd^4} \qquad \text{for round wire } (J = 2I)$$

(b) Rotation prevented:
The restraint of rotation introduces a force along the wire. The incoming load

and torque give a torque $\frac{1}{2}PD\cos\theta$ and a bending moment $\frac{1}{2}PD\sin\theta$

$$\text{The strain energy} = (\tfrac{1}{2}PD\cos\theta)^2\,\frac{L}{2GJ} + (\tfrac{1}{2}PD\sin\theta)^2\,\frac{L}{2EI}$$

Using $E = 2(1+v)G$ and $J = 2I$ as above, and $\cos^2\theta = 1 - \sin^2\theta$

$$\frac{dh}{dP} = \frac{8nD^3\sec\theta\left(1 - \dfrac{v}{1+v}\sin^2\theta\right)}{Gd^4} \qquad \text{for round wire}$$

A.8 BEAMS WITH RESILIENT SUPPORTS

Figure A.9a shows a beam with a rolling load P and a central support which is much more resilient than the end supports. Its stiffness is k units of force per unit of deflection. We wish to find the highest bending moments as P rolls from end to end. Peak values occur at P and R.

We find the reaction R by deflection considerations, then we can find Q and S by taking moments. The bending moments are Qx and $SL/2$.

By Maxwell (section 7.11) the deflection at the centre due to a load P at x can be taken from the standard case for a central load, figure 7.3. Calling this Δ_P

$$\Delta_P = \frac{P\left(\dfrac{xL^2}{16} - \dfrac{x^3}{12}\right)}{EI}$$

The deflection due to R alone is Δ_R

$$\Delta_R = \frac{RL^3}{48EI}$$

From the stiffness

$$R = K(\Delta_p - \Delta_R)$$

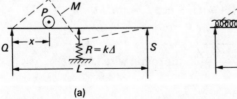
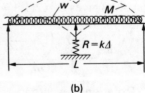

(a)   (b)

Figure A.9

If the end supports were also resilient, we should require additional equations at this stage, involving Q and S.

Substituting for the deflections

$$\frac{R}{k} = \frac{P\left(\frac{xL^2}{16} - \frac{x^3}{12}\right)}{EI - \frac{RL^3}{48EI}}$$

Rearranging

$$\frac{R}{P} = \frac{\left(\frac{xL^2}{16} - \frac{x^3}{12}\right)}{\left(\frac{L^3}{48} + \frac{EI}{k}\right)} \tag{A.6}$$

By moments

$$Q = P\left(1 - \frac{x}{L}\right) - \frac{R}{2} \quad S = \frac{Px}{L} - \frac{R}{2} \text{ (may be positive or negative)}$$

$$M_P = Qx = Px - \frac{Px^2}{L} - \frac{Rx}{2}$$

$$\frac{M_P}{P} = x - \frac{x^2}{L} - \frac{\left(\frac{x^2 L^2}{32} - \frac{x^4}{24}\right)}{\left(\frac{L^3}{48} + \frac{EI}{k}\right)} \tag{A.7}$$

For a maximum, $\mathrm{d}M/\mathrm{d}x = 0$

$$1 - \frac{2x}{L} = \frac{\left(\frac{xL^2}{16} - \frac{x^3}{6}\right)}{\left(\frac{L^3}{48} + \frac{EI}{k}\right)} \tag{A.8}$$

This gives a cubic equation in x, to be solved by trial and error. If k is small, $x \rightarrow L/2$; if k is large, $x \approx L/4$. The x value is then substituted into equation A.7 to give the moment at P. The moment at R is generally smaller.

With a UDL the solution is much simpler. The deflections are found from figure 7.3. Then

$$\Delta_w - \Delta_R = \frac{5wL^4}{384EI} - \frac{RL^3}{48EI} = \frac{R}{k}$$

Rearranging

$$R = \frac{5wL^4}{384\left(\frac{L^3}{48} + \frac{EI}{k}\right)} \tag{A.9}$$

The bending moment may be greatest at the centre, which is a discontinuity and will not show up by differentiating, or somewhere near $L/4$. We shall not proceed beyond writing the bending moment equation since after that the process is similar to the above but including an independent check on the central bending moment, obtained by putting $x = L/2$.

$$M = \frac{w(Lx - x^2)}{2} - \frac{Rx}{2} \qquad (A.10)$$

When the resilient support is continuous and the beam has concentrated loading or a local moment applied to it we have the classical case of a beam on an elastic foundation. It may be thought that this applies to railway tracks; unfortunately the support due to the ballast and earth behaves unsymmetrically, giving zero force when the deflection is upwards. In the downward direction it is also liable to be non-linear. However, there are engineering applications as we have seen in sections 7.10 (second part) and 9.5.

We only consider long beams, that is, beams so long in relation to the stiffnesses involved that the deflections fade away to negligible values towards the end of the beam. The general case is the semi-infinite beam with a force and a couple at $x = 0$, figure A.10. Following the usual notation for this topic force and deflection are positive downwards. The stiffness of the 'ground' is k units of force per unit length per unit deflection.

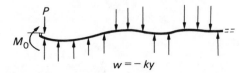

Figure A.10

The type of 'ground' we are interested in reacts linearly to deflection and also gives downward restraint when the deflection is above base-line. This entitles us to use equation 7.7, the loading w being proportional to y

$$-w = EI\, d^4y/dx^4 = ky \qquad (A.11)$$

(See section 7.1 about the sign convention problem.) The solution is found in various textbooks, in this form or rearranged

$$y = \frac{e^{-mx}[P\cos mx - mM_0\,(\cos mx - \sin mx)]}{2m^3 EI} \qquad (A.12)\dagger$$

† In early editions of reference 5 the statement equivalent to equation A.12 is dimensionally inconsistent in powers of m, giving m^2 instead of m^3 in the denominator. Third and later editions are corrected.

$$M = e^{-mx}[M_0(\cos mx + \sin mx) - \frac{P}{m} \sin mx] \qquad (A.13)$$

$m = (k/4EI)^{1/4}$; by dimensional analysis we note that m is the reciprocal of a length. The length $1/m$ is a characteristic length of the deflected shape, both in terms of decay rate and of wavelength.

A.9 SOLID CURVED BARS

Following figure A.11, line OA remains undeflected. Under applied forces and moments a line OZ originally at an angle θ is deflected by $d\theta$, with a neutral point located at $x = h$ (as yet unknown) where x is a general radius variable.

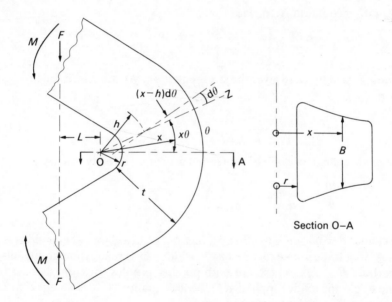

Figure A.11

At x, an original length $x\theta$ is extended by $(x - h)\, d\theta$; $\sigma = E \times$ strain

$$\sigma = E\, d\theta \frac{(x - h)}{x\theta} = \frac{E\, d\theta}{\theta}\left(1 - \frac{h}{x}\right) \qquad (A.14)$$

$$\text{Force } F = \int_r^{r+t} \sigma B\, dx = \frac{E\, d\theta}{\theta} \int_r^{r+t} B\left(1 - \frac{h}{x}\right) dx \qquad (A.15)$$

Moments about O

$$M + FL = \int_r^{r+t} \sigma B x \, dx = \frac{E \, d\theta}{\theta} \int_r^{r+t} B \, (x - h) \, dx \qquad \text{(A.16)}$$

Note that the only assumption made is linearity of stress with distance from the neutral axis. This is not the requirement of plane sections remaining plane; distortion due to shear, etc., is accepted provided it does not violate the linearity (see figure 2.27).

Solution for constant width, pure moment.

B = constant; from equation A.15, integrated and set to zero

$$r + t - r - \frac{h \ln (r + t)}{r} = 0$$

$$h = \frac{t}{\ln (1 + t/r)} \qquad \text{(A.17)}$$

Integrating equation A.16, with $F = 0$

$$M = \frac{E \, d\theta}{\theta} B \left[\frac{x^2}{2} - hx \right]_r^{r+t} = \frac{E \, d\theta}{\theta} B \left(rt - ht + \frac{t^2}{2} \right) \qquad \text{(A.18)}$$

Dividing A.14 by A.18 gives the stress equation. We are interested in σ_{max}, at $x = r$

$$\sigma_{max} = \frac{M(h/r - 1)}{B(rt - ht + t^2/2)} \qquad \text{(A.19)}$$

The angle change per unit of subtended arc comes directly from A.18

$$\frac{d\theta}{\theta} = \frac{M}{EB} \left(rt - ht + \frac{t^2}{2} \right) \qquad \text{(A.20)}$$

We cannot devote space to other profiles but from a glance at equations A.15 and A.16 a trapezium expressed as $B = B_0 - kx$ is readily integrable, also a hyperbola $B = k/x$. These are both popular profiles for crane hooks. In a crane hook, the load is applied at O, hence equation A.16 is set equal to zero which simplifies the mathematics.

These equations were used to give the data in section 7.10. They agree with some of the previously published data.

To illustrate the case, we take an example. Figure A.12 shows a clamp, 20 mm wide, with a force of 1 kN as shown. Find the highest stress.

Force equation

$$1000 = \frac{E \, d\theta}{\theta} \int_8^{16} 20 \left(1 - \frac{h}{x} \right) dx = \frac{E \, d\theta}{\theta} \times 20 \, (8 - h \ln 2)$$

Moment equation about O

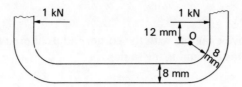

Figure A.12

$$1000 \times 12 = -\frac{E\,\mathrm{d}\theta}{\theta} \int_8^{16} 20\,(x - h)\,\mathrm{d}x$$

$$= -\frac{E\,\mathrm{d}\theta}{\theta} \times 20 \times (96 - 8h)$$

(Note sign to give same sign for stresses as in force equation.)
Divide

$$96 - 8h = -12(8 - 0.693h)$$

$$192 = 16.316h$$

$$h = 11.76\,\mathrm{mm}$$

Maximum stress is at $x = r = 8$ mm

$$\sigma = \frac{E\,\mathrm{d}\theta}{\theta}\left(1 - \frac{11.76}{8}\right)$$

Find $E\,\mathrm{d}\theta/\theta$ from force equation

$$\frac{1000}{20(8 - 11.76 \times 0.693)} = -334$$

$$\sigma = 157\,\mathrm{N/mm^2}$$

For interest, compare this with the superposition answer using the SCF from
section 7.10

$$\text{Stress due to moment at mean line} = \frac{6M}{Bt^2} = \frac{6 \times 1000 \times 24}{20 \times 64} =$$

$$112.5\,\mathrm{N/mm^2}$$

$$\text{SCF} = 1.29$$

$$\text{direct stress} = \frac{1000}{8 \times 20} = 6.25\,\mathrm{N/mm^2}$$

$$\text{Total} = 6.25 + 1.29 \times 112.5 = 151\,\mathrm{N/mm^2}$$

Problem A.1

Repeat the above example with the corner radius altered to 2 mm, all other
stated dimensions remaining the same.

Ans. 167 N/mm².

Comments:

(1) although the new form has a smaller bending moment, the stress is
higher;

(2) answer by superposition of pure moment and SCF plus direct stress
gives 187 N/mm². The first answer is right.

A.10 THE CONNECTION BETWEEN THE ELASTIC CONSTANTS E, G AND ν

The simple tension situation with its attendant shear gives rise to a particularly
short demonstration of the relation between the elastic moduli and Poisson's
ratio, namely $E = 2(1 + \nu)G$.

Figure A.13 shows an element of material unstressed (solid lines) and with
a tensile stress σ (chain-dotted).

$$\text{The longitudinal extension } \Delta = \frac{\sigma a}{E} \tag{A.21}$$

the transverse contraction $= \nu\Delta$ by definition of ν.

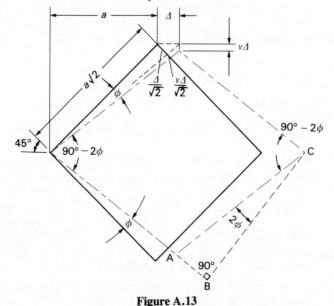

Figure A.13

If the basic angle is 45° as shown and angle changes are small, then from the figure

$$\frac{\left(\dfrac{\Delta}{\sqrt{2}} + \dfrac{\nu\Delta}{\sqrt{2}}\right)}{a\sqrt{2}} = \phi \tag{A.22}$$

To find the shear angle construct the right-angled triangle ABC. This shows tha the shear angle of the stressed shape is 2ϕ, $G = \tau/2\phi$ (definition of G).

Having shown that in simple tension the shear stress at 45° is half the normal tensile stress (equation 2.31)

$$\tau = \frac{\sigma}{2}$$

Therefore

$$\frac{(1 + \nu)\Delta}{2a} = \frac{\tau}{2G} \quad \text{(from A.22)}$$

Substituting for Δ from equation A.21 and cancelling a

$$\frac{(1 + \nu)\sigma}{2E} = \frac{\tau}{2G} = \frac{\sigma}{4G}$$

Rearranging

$$E = 2(1 + \nu)G \tag{A.23}$$

This treatment is readily extendable to include the bulk modulus (isothermal compressibility) but this is of little engineering interest because of triaxial stresses commonly present. These require further equations.

A.11 CONSTRUCTION OF MOHR'S CIRCLE FOR STRESS GIVEN σ_x, σ_y AND τ_{xy}

Note that σ_y is not yield stress in this section but tensile stress in the y dir. direction. Use the x axis for both tensile stresses, σ_x and σ_y. The y axis is used for the shear stress τ. It is essential to use the same scales along x and y.

Plot the points σ_x, τ and σ_y, $-\tau$. Compressive stresses are treated as negative tensions. Join these points and draw a circle using the joining line as diameter.

Figure A.14 shows two examples, one with σ_y positive, one negative. All possible stress combinations lie on the circle. The maxima are found as indicated. Below these, two examples are shown with $\sigma_y = 0$, identical σ_x but different τ values.

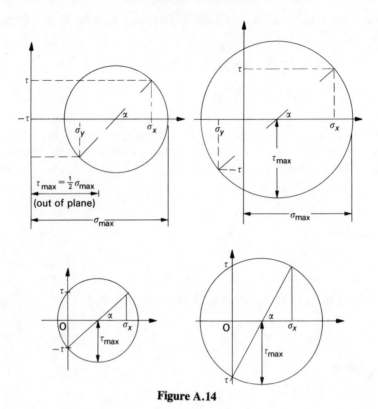

Figure A.14

An additional item obtainable from Mohr's circle is the angle between the x axis and the direction of σ_{max}. It is $\alpha/2$. The maximum shear stress is at 45° to the maximum tensile stress. Knowledge of these angles is useful in interpreting strain-gauge results.

A.12 FUNICULAR POLYGON FOR BENDING MOMENTS

This is a graphical routine for finding the bending moment along a beam with many loads. It is not only useful in itself but immediately leads to a solution for built-in beams, by moment area. The procedure is shown by examples. In figure A.15a we set out the loads vertically as in Bow's notation procedure. Horizontal components are ignored since they do not produce bending moments; any actual sloping loads must be resolved vertically at the neutral axis. Then we choose an arbitrary pole P and draw lines parallel to the rays Pa,

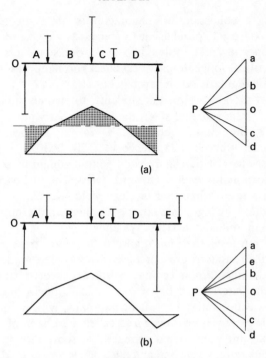

Figure A.15

Pb, etc., on to the vertical lines corresponding with the load positions. It is an advantage to set the pole level with o and make the pole distance a convenient unit value. Then the bending moment at any point can be read off directly as the vertical height. If the beam is built-in, we simply put in a raised base-line until the moment areas above and below are equal (shown shaded).

The second example (figure A.15b) shows an overhung load treated by the funicular polygon; in this instance there is no question of built-in ends.

Provided that there are no overhanging loads, we can draw the diagram easily without calculating the reactions. In general, the first and last points will fall on different levels. By joining them up and drawing a line parallel to this, through P on to the vertical axis, we obtain the reactions *do*, *oa*. With overhangs, the procedure is a little more complicated, it may be found in reference 72. However, in such cases the graphical method may be no better than calculation.

A.13 A SHORT NOTE ON TYPICAL MATERIAL PROPERTIES

An attempt at accurate stress calculations, even if only to ± 10 per cent, would be pointless without reasonably accurate materials data. A design office generally has a collection of manufacturers' data sheets, national and company standards. If purchasing to standards we can rely on the stated minimum properties, preferably supported by occasional tests on the material received. In critical applications, works checks on composition and properties would be made regularly on all new batches.

The present section is aimed at students, to give an outline of the likely ranges. In the great range of steels the only property more or less common throughout is the elastic behaviour as far as it goes; E is close to 200 000 N/mm², ν about 0.28, hence G is $200\,000/2(1 + 0.28) = 78\,000$ N/mm². Those steels with a sharp yield point maintain these properties to a limit of proportionality well up towards yield. The high-strength materials with a proof stress rather than a clear yield will depart from constants gradually. If there are high residual stresses, there may be apparent departure from Young's modulus, due to local yielding.

For design purposes the main figure is the yield stress σ_y. In structural steelwork, rolled sections are purchased by specified grades of UTS, with specified yield points. When it comes to ordinary sheet or bar from the stores, fully annealed material can be of $\sigma_y \approx 120$ N/mm² while rolled bars or bright-drawn sections may be soft or half-hard, with σ_y up to 250 to 300 N/mm² in mild steel. There is even a so-called mild steel, of less than 0.2 per cent carbon but fine-grained and stronger than some alloy steels. Then there is a tremendous range of carbon and alloy steels. We should remember that the properties of any one mix depend also on history, that is, work-hardening, heat-treatment and ageing, on the ruling thickness (by way of the grain size) and on the direction of test.

Aluminium and its alloys also cover a very wide band of properties, with a common E value of around 70 000 N/mm², yield point or proof stress ranging from 30 N/mm² for annealed pure aluminium to 500 N/mm² for the strong alloys. The low E and high strength combined with the methods of heat-treatment give rise to greater distortion than many other metals, during machining or heat-treatment.

The copper, brass and bronze system is more variable; large amounts of zinc or aluminium are used in some compositions so that E can range from 100 000 for brass to 125 000 for copper and many bronzes; it varies somewhat with temperature. The strength starts nearly as low as for aluminium but goes up twice as high for aluminium- and beryllium-bearing alloys.

Unusually low E values are shown by magnesium alloys, zinc and the low melting-point metals. Anomalous elastic behaviour occurs in flake cast iron; due to the opening-up effect of the network of graphite flakes, E in tension is significantly lower than in compression, to an extent varying considerably with composition. This affects the bending properties also.

Concrete varies substantially in properties with the mix and possibly with age. The main point is that its E is 7 to 15 times smaller than steel, which has implications in the placing or reinforcement in concrete structures. The discrepancy demands good bonding and good end-fixings.

A similar discrepancy exists between glass or carbon fibres and the resins with which they are bonded.

Plastics properties are somewhat more difficult to define than metal properties. The strength is time-dependent in many to an extent comparable with metals at high temperature; the elastic properties are stress-dependent so that for accurate calculations it may be necessary to define Young's modulus as the secant modulus between a given stress level and zero, or the tangent modulus for small changes.

For metal properties, Smithell's *Metals Reference Book* is suggested. [73] The short table below is not exhaustive, nor guaranteed correct. It is merely a rough guide for students to help with a realistic appreciation.

A.14 SELECTED STRUCTURAL SECTIONS

A condensed version of the main properties of some rolled steel sections normally available in the United Kingdom is given below, extracted and condensed from the BCSA handbook, [48] by permission. U.S. sections cover a similar range but with at least double the choice of thicknesses, also some additional overall ratios. The hollow rectangular section range is particularly generous, offering sections such as 12 in × 4 in, 12 in × 2 in, 8 in × 3 in.

To save space the radii of gyration, elastic and plastic moduli have been omitted, also the fillet radii. Note the inclusion of a net *I* value, allowing for bolt-holes. This would be used for stress calculations when holes are required in a region of high stress but would be ignored for deflection purposes unless there are a great many holes.

Joists are the traditional *I*-beams, their flanges tapered at 5 or 8° (see footnote on table); when bolting these, taper washers of the correct form should be used to avoid bending the bolts. Universal beams and columns take their name from the universal mill with four rolls, able to produce a choice of profiles from the same set of rolls (figure A.16). Other omissions are

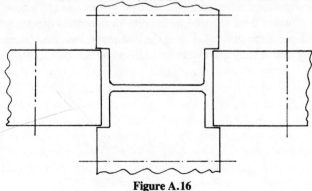

Figure A.16

channels, angles, tees, bulb bars and bulb angles used in shipbuilding (to stiffen the hull while giving good access for welding).

Angle sections are available in a large range of widths and thicknesses, from $25 \times 25 \times 3$ mm to $200 \times 200 \times 24$ mm, including various unequal-legged angles. One important use is as struts, hence is is useful to be reminded that their lowest I value is on an oblique axis. For buckling calculations the following short rule is possible.

Equal angles, external measurement $A \times A$, lowest radius of gyration = $A/5.2$. Unequal angles, external measurement $A \times B$, B is the shorter leg, lowest radius of gyration = $B/4.7$.

Some notes in the tables relate to reference 48 and may be ignored for present purposes.

A.15 THIN-WALL BUCKLING

(a) Thin-walled cylinders
Thin-walled cylinders can buckle under axial compression or under axial compressive stress caused by a bending moment. Cylinders of wall thickness less than $D/800$ may buckle under axial compressive stresses above 0.2 Et/D where t is the wall thickness, D is the diameter, see figure 2 of Hart-Smith [75]. Thicker cylinders may start to buckle when the compressive stress exceeds 1.2 Et/D. An initial crease greatly lowers the threshold; it may be halved if the crease is $\frac{1}{2}t$ deep. Treat as in section 2.3.3 if stress is near to yield point.

(b) Dished end covers
A number of recent papers have dealt with creasing of very thin dished end covers. This work may be of interest in such vessels as come outside the requirements of the pressure vessel codes, for example, some forms of food-processing plant or storage tanks [76, 77].

(c) Plate girders
Recent investigations have concentrated on girder panels whose webs are appreciably thinner than allowed by current design rules in structural codes; also with end-posts much weaker than the flanges and without flange angles or tongue plates. Since they are rather specialised no references are given here.

Outline Guide to Material Properties

Strength properties vary widely by thickness, state of cold work or heat-treatment

Approximate description of material type. Details reference 73, makers' data, BS, DTD or US standards, etc.		UTS (N/mm²)	σ_y or proof stress (N/mm²)	E (MN/mm²)	ν	α (mm/mm °C × 10⁶)	ρ (g/cm³)
Mild steel 0.1% C (En1, SAE 1010) annealed		220	140	0.2	0.28	11.5	7.8
bright drawn		350	240	0.2	0.28	11.5	7.8
Mild structural steel 0.2% C, as rolled		430	250	0.21	0.28	11.5	7.8
Strong structural steel (requires preheat for welding)		550	420	0.21	0.28	11.5	7.8
High-tensile steels, heat-treated for strength		up to 1500		0.21	0.3	11.5	7.8
Super-alloy (steels)		up to 2200					
Cr–Ni stainless steels 18/8 + stabilisers		300	550	0.215	0.28	13–16	≈7.7
Al–4.5% Mg alloy as-rolled, extruded (N8)		300–370	170–290	0.07	0.35	23	2.7
Al–11% Si alloy, sand-cast		180	130	0.07	0.35	20	2.7
Al–Mn, etc., alloy, weldable, extruded (H20)		300	250	0.07	0.35	23	2.7
Al–Cu– etc. non-weldable, heat-treated for strength	up to	450 600	400 500	0.07	0.35	22	2.8
Mg–Al alloys, sand-cast		200	120	0.05	0.35	26	1.8
65/35 Cu–Zn brass, soft		80	50	0.1	0.3	19	8.5
as-rolled or ½ hard		180	100	0.1	0.3	19	8.5
Gunmetal (Bronze) cast 88 Cu, 10 Sb, 2 Zn		300	150	0.12	0.3	17	8.8
High-tensile propeller bronze, cast	up to	700	400	0.12	0.33	18	8.8
Diecast zinc alloy		300	220	0.05	0.3	26	6.5
Titanium 318 alloy		1000	950	0.11	0.33	8	4.5
Titanium 155 alloy		660	530	0.11	0.33	8	4.5

UNIVERSAL BEAMS

DIMENSIONS AND PROPERTIES

Serial size	Mass per metre	Depth of section D	Width of section B	Thickness Web t	Thickness Flange T	Moment of inertia Axis x-x Gross	Moment of inertia Axis x-x * Net	Moment of inertia Axis y-y	Area of section
mm	kg	mm	mm	mm	mm	cm⁴	cm⁴	cm⁴	cm²
914 × 419	388	920.5	420.5	21.5	36.6	717325	642716	42481	493.9
	343	911.4	418.5	19.4	32.0	623866	559325	36251	436.9
914 × 305	289	926.6	307.8	19.6	32.0	503781	437962	14793	368.5
	253	918.5	305.5	17.3	27.9	435796	379081	12512	322.5
	224	910.3	304.1	15.9	23.9	375111	327298	10425	284.9
	201	903.0	303.4	15.2	20.2	324715	284809	8632	256.1
838 × 292	226	850.9	293.8	16.1	26.8	339130	315154	10661	288.4
	194	840.7	292.4	14.7	21.7	278833	259610	8384	246.9
	176	834.9	291.6	14.0	18.8	245412	228869	7111	223.8
762 × 267	197	769.6	268.0	15.6	25.4	239464	221138	7699	250.5
	173	762.0	266.7	14.3	21.6	204747	189347	6376	220.2
	147	753.9	265.3	12.9	17.5	168535	156195	5002	187.8
686 × 254	170	692.9	255.8	14.5	23.7	169843	156120	6225	216.3
	152	687.6	254.5	13.2	21.0	150015	137962	5391	193.6
	140	683.5	253.7	12.4	19.0	135972	125170	4789	178.4
	125	677.9	253.0	11.7	16.2	117700	108590	3992	159.4
914 × 419	388	920.5	420.5	21.5	36.6	718742	644135	45407	494.5
	343	911.4	418.5	19.4	32.0	625282	560743	39150	437.5
914 × 305	289	926.6	307.8	19.6	32.0	504594	438776	15610	368.8
	253	918.5	305.5	17.3	27.9	436610	379896	13318	322.8
	224	910.3	304.1	15.9	23.9	375924	328112	11223	285.3
	201	903.0	303.4	15.2	20.2	325529	285623	9427	256.4
838 × 292	226	850.9	293.8	16.1	26.8	339747	315771	11353	288.7
	194	840.7	292.4	14.7	21.7	279450	260228	9069	247.2
	176	834.9	291.6	14.0	18.8	246029	229487	7792	224.1

762 × 267	197	769.6	268.0	15.6	25.4	239894	221568	8174	250.8
	173	762.0	266.7	14.3	21.6	205177	189778	6846	220.5
	147	753.9	265.3	12.9	17.5	168966	156626	5468	188.1
686 × 254	170	692.9	255.8	14.5	23.7	170147	156424	6621	216.6
	152	687.6	254.5	13.2	21.0	150319	138266	5782	193.8
	140	683.5	253.7	12.4	19.0	136276	125474	5179	178.6
	125	677.9	253.0	11.7	16.2	118003	108894	4379	159.6
610 × 305	238	633.0	311.5	18.6	31.4	207571	178026	15838	303.8
	179	617.5	307.0	14.1	23.6	151631	129960	11412	227.9
	149	609.6	304.8	11.9	19.7	124660	106842	9300	190.1
610 × 229	140	617.0	230.1	13.1	22.1	111844	101652	4512	178.4
	125	611.9	229.0	11.9	19.6	98579	89634	3933	159.6
	113	607.3	228.2	11.2	17.3	87431	79590	3439	144.5
	101	602.2	227.6	10.6	14.8	75720	69087	2912	129.2
533 × 210	122	544.6	211.9	12.8	21.3	76207	68609	3393	155.8
	109	539.5	210.7	11.6	18.8	66739	60111	2937	138.6
	101	536.7	210.1	10.9	17.4	61659	55557	2694	129.3
	92	533.1	209.3	10.2	15.6	55353	49912	2392	117.8
	82	528.3	208.7	9.6	13.2	47491	42934	2005	104.4

Values in the shaded area relate to Universal Beams with tapered flanges.

(OBSOLESCENT)

* WITH HOLES

* Section	Serial or nominal flange width (mm)	Holes (in each flange)	
		Number	Diameter (mm)
Universal beams	305 to 419	2	26
	165 to 292	1	26
Universal columns	114 to 152	1	22
and	76 to 102	1	14
joists	64	1	12
	44 and 51	1	10
Channels	102	1	26
	76 and 89	1	22
	64	1	18
	38 and 51	1	12

Note. In some cases the size of hole deducted is excessive in respect of the minimum edge distance specified in BS 449, Table 21.

UNIVERSAL BEAMS
CONT.

DIMENSIONS AND PROPERTIES

Serial size	Mass per metre	Depth of section D	Width of section B	Thickness Web t	Thickness Flange T	Moment of inertia Axis x–x Gross	* Net	Axis y–y	Area of section
mm	kg	mm	mm	mm	mm	cm⁴	cm⁴	cm⁴	cm²
457 × 191	98	467.4	192.8	11.4	19.6	45717	40615	2343	125.3
	89	463.6	192.0	10.6	17.7	41021	36456	2086	113.9
	82	460.2	191.3	9.9	16.0	37103	32996	1871	104.5
	74	457.2	190.5	9.1	14.5	33388	29698	1671	95.0
	67	453.6	189.9	8.5	12.7	29401	26190	1452	85.4
457 × 152	82	465.1	153.5	10.7	18.9	36215	32074	1143	104.5
	74	461.3	152.7	9.9	17.0	32435	28744	1012	95.0
	67	457.2	151.9	9.1	15.0	28577	25357	878	85.4
	60	454.7	152.9	8.0	13.3	25464	22611	794	75.9
	52	449.8	152.4	7.6	10.9	21345	19035	645	66.5
406 × 178	74	412.8	179.7	9.7	16.0	27329	24062	1545	95.0
	67	409.4	178.8	8.8	14.3	24329	21425	1365	85.5
	60	406.4	177.8	7.8	12.8	21508	18934	1199	76.0
	54	402.6	177.6	7.6	10.9	18626	16457	1017	68.4
406 × 140	46	402.3	142.4	6.9	11.2	15647	13765	539	59.0
	39	397.3	141.8	6.3	8.6	12452	11017	411	49.4
356 × 171	67	364.0	173.2	9.1	15.7	19522	17045	1362	85.4
	57	358.6	172.1	8.0	13.0	16077	14053	1109	72.2
	51	355.6	171.5	7.3	11.5	14156	12384	968	64.6
	45	352.0	171.0	6.9	9.7	12091	10609	812	57.0
356 × 127	39	352.8	126.0	6.5	10.7	10087	9213	357	49.4
	33	348.5	125.4	5.9	8.5	8200	7511	280	41.8

305 × 165	54	310.9	166.8	7.7	13.7	11710	10134	1061	68.4
	46	307.1	165.7	6.7	11.8	9948	8609	897	58.9
	40	303.8	165.1	6.1	10.2	8523	7384	763	51.5
305 × 127	48	310.4	125.2	8.9	14.0	9504	8643	460	60.8
	42	306.6	124.3	8.0	12.1	8143	7409	388	53.2
	37	303.8	123.5	7.2	10.7	7162	6519	337	47.5
305 × 102	33	312.7	102.4	6.6	10.8	6487	5800	193	41.8
	28	308.9	101.9	6.1	8.9	5421	4862	157	36.3
	25	304.8	101.6	5.8	6.8	4387	3962	120	31.4
254 × 146	43	259.6	147.3	7.3	12.7	6558	5706	677	55.1
	37	256.0	146.4	6.4	10.9	5556	4834	571	47.5
	31	251.5	146.1	6.1	8.6	4439	3879	449	40.0
254 × 102	28	260.4	102.1	6.4	10.0	4008	3569	178	36.2
	25	257.0	101.9	6.1	8.4	3408	3046	148	32.2
	22	254.0	101.6	5.8	6.8	2867	2575	120	28.4
203 × 133	30	206.8	133.8	6.3	9.6	2887	2476	384	38.0
	25	203.2	133.4	5.8	7.8	2356	2027	310	32.3

* see p. 293

JOISTS

DIMENSIONS AND PROPERTIES

Nominal size	Mass per metre	Depth of section D	Width of section B	Thickness Web t	Thickness Flange T	Moment of inertia Axis x-x Gross	Moment of inertia Axis x-x *Net	Moment of inertia Axis y-y	Area of section
mm	kg	mm	mm	mm	mm	cm⁴	cm⁴	cm⁴	cm²
254 × 203	81.85	254.0	203.2	10.2	19.9	12016	10527	2278	104.4
254 × 114	37.20	254.0	114.3	7.6	12.8	5092	4243	270.1	47.4
203 × 152	52.09	203.2	152.4	8.9	16.5	4789	4177	813.3	66.4
203 × 102	25.33	203.2	101.6	5.8	10.4	2294	2024	162.6	32.3
178 × 102	21.54	177.8	101.6	5.3	9.0	1519	1339	139.2	27.4
152 × 127	37.20	152.4	127.0	10.4	13.2	1818	1627	378.8	47.5
152 × 89	17.09	152.4	88.9	4.9	8.3	881.1	762.6	85.98	21.8
152 × 76	17.86	152.4	76.2	5.8	9.6	873.7	736.2	60.77	22.8
127 × 114	29.76	127.0	114.3	10.2	11.5	979.0	866.9	241.9	37.3
127 × 114	26.79	127.0	114.3	7.4	11.4	944.8	834.6	235.4	34.1
127 × 76	16.37	127.0	76.2	5.6	9.6	569.4	476.1	60.35	21.0
127 × 76	13.36	127.0	76.2	4.5	7.6	475.9	400.0	50.18	17.0

114 × 114	26.79	114.3	114.3	9.5	10.7	735.4	651.2	223.1	34.4
102 × 102	23.07	101.6	101.6	9.5	10.3	486.1	425.1	154.4	29.4
102 × 64	9.65	101.6	63.5	4.1	6.6	217.6	182.2	25.30	12.3
102 × 44	7.44	101.6	44.4	4.3	6.1	152.3	126.9	7.91	9.5
89 × 89	19.35	88.9	88.9	9.5	9.9	306.7	263.7	101.1	24.9
76 × 76	14.67	76.2	80.0	8.9	8.4	171.9	144.1	60.77	19.1
76 × 76	12.65	76.2	76.2	5.1	8.4	158.6	130.7	52.03	16.3

Sections with mass shown in italics are, although frequently rolled, not in BS 4. Availability should be checked with BSC Sections Product Unit. Flanges of BS 4 joists have a 5° taper; all others taper at 8°.

* see p. 293

UNIVERSAL COLUMNS

DIMENSIONS AND PROPERTIES

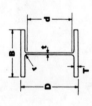

Serial size	Mass per metre	Depth of section D	Width of section B	Thickness Web t	Thickness Flange T	Moment of inertia Axis x–x Gross	Moment of inertia Axis x–x *Net	Moment of inertia Axis y–y	Area of section
mm	kg	mm	mm	mm	mm	cm⁴	cm⁴	cm⁴	cm²
356 × 406	634	474.7	424.1	47.6	77.0	275140	243065	98211	808.1
	551	455.7	418.5	42.0	67.5	227023	200308	82665	701.8
	467	436.6	412.4	35.9	58.0	183118	161334	67905	595.5
	393	419.1	407.0	30.6	49.2	146765	129152	55410	500.9
	340	406.4	403.0	26.5	42.9	122474	107674	46816	432.7
	287	393.7	399.0	22.6	36.5	99994	87837	38714	366.0
	235	381.0	395.0	18.5	30.2	79110	69430	31008	299.8
Column Core	477	427.0	424.4	48.0	53.2	172391	152946	68056	607.2
356 × 368	202	374.7	374.4	16.8	27.0	66307	57805	23632	257.9
	177	368.3	372.1	14.5	23.8	57153	49791	20470	225.7
	153	362.0	370.2	12.6	20.7	48525	42263	17469	195.2
	129	355.6	368.3	10.7	17.5	40246	35047	14555	164.9
305 × 305	283	365.3	321.8	26.9	44.1	78777	66878	24545	360.4
	240	352.6	317.9	23.0	37.7	64177	54405	20239	305.6
	198	339.9	314.1	19.2	31.4	50832	43039	16230	252.3
	158	327.2	310.6	15.7	25.0	38740	32783	12524	201.2
	137	320.5	308.7	13.8	21.7	32838	27782	10672	174.6
	118	314.5	306.8	11.9	18.7	27601	23344	9006	149.8
	97	307.8	304.8	9.9	15.4	22202	18776	7268	123.3

254 × 254	167	289.1	264.5	19.2	31.7	29914	27171	9796	212.4
	132	276.4	261.0	15.6	25.3	22575	20492	7519	168.9
	107	266.7	258.3	13.0	20.5	17510	15889	5901	136.6
	89	260.4	255.9	10.5	17.3	14307	12973	4849	114.0
	73	254.0	254.0	8.6	14.2	11360	10299	3873	92.9
203 × 203	86	222.3	208.8	13.0	20.5	9462	8373	3119	110.1
	71	215.9	206.2	10.3	17.3	7647	6756	2536	91.1
	60	209.6	205.2	9.3	14.2	6088	5383	2041	75.8
	52	206.2	203.9	8.0	12.5	5263	4651	1770	66.4
	46	203.2	203.2	7.3	11.0	4564	4035	1539	58.8
152 × 152	37	161.8	154.4	8.1	11.5	2218	1931	709	47.4
	30	157.5	152.9	6.6	9.4	1742	1516	558	38.2
	23	152.4	152.4	6.1	6.8	1263	1104	403	29.8

* see p. 293

SQUARE HOLLOW SECTIONS

DIMENSIONS AND PROPERTIES

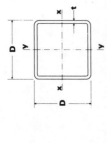

| Designation | | Mass per metre | Area of section | Moment of inertia | Torsional constants | |
| Size D × D | Thickness t | | | | J | C |
mm	mm	kg	cm²	cm⁴	cm⁴	cm³
20 × 20	2.0	1.12	1.42	0.76	1.22	1.07
	2.6	1.39	1.78	0.88	1.44	1.23
30 × 30	2.6	2.21	2.82	3.49	5.56	3.30
	3.2	2.65	3.38	4.00	6.45	3.75
40 × 40	2.6	3.03	3.86	8.94	14.0	6.41
	3.2	3.66	4.66	10.4	16.5	7.43
	4.0	4.46	5.68	12.1	19.5	8.56
50 × 50	3.2	4.66	5.94	21.6	33.8	12.4
	4.0	5.72	7.28	25.5	40.4	14.5
	5.0	6.97	8.88	29.6	47.6	16.7
60 × 60	3.2	5.67	7.22	38.7	60.1	18.6
	4.0	6.97	8.88	46.1	72.4	22.1
	5.0	8.54	10.9	54.4	86.3	25.8
70 × 70	3.6	7.46	9.50	69.5	108	28.7
	5.0	10.1	12.9	90.1	142	36 8

| Designation | | Mass per metre | Area of section | Moment of inertia | Torsional constants | |
| Size D × D | Thickness t | | | | J | C |
mm	mm	kg	cm²	cm⁴	cm⁴	cm³
150 × 150	5.0	22.7	28.9	1009	1548	197
	6.3	28.3	36.0	1236	1907	240
	8.0	35.4	45.1	1510	2348	291
	10.0	43.6	55.5	1803	2829	345
	12.5	53.4	68.0	2125	3372	403
	16.0	66.4	84.5	2500	4029	468
180 × 180	6.3	34.2	43.6	2186	3357	355
	8.0	43.0	54.7	2689	4156	434
	10.0	53.0	67.5	3237	5041	519
	12.5	65.2	83.0	3856	6062	613
	16.0	81.4	104	4607	7339	725
200 × 200	6.3	38.2	48.6	3033	4647	444
	8.0	48.0	61.1	3744	5770	545
	10.0	59.3	75.5	4525	7020	655
	12.5	73.0	93.0	5419	8479	779
	16.0	91.5	117	6524	10330	929

250 × 250	6.3	48.1	61.2	6049	9228	712
	8.0	60.5	77.1	7510	11511	880
	10.0	75.0	95.5	9141	14086	1065
	12.5	92.6	118	11050	17139	1279
	16.0	117	149	13480	21109	1548
300 × 300	10.0	90.7	116	16150	24776	1575
	12.5	112	143	19630	30290	1905
	16.0	142	181	24160	37566	2327
350 × 350	10.0	106	136	26050	39840	2186
	12.5	132	168	31810	48869	2655
	16.0	167	213	39370	60901	3265
400 × 400	10.0	122	156	39350	60028	2896
	12.5	152	193	48190	73815	3530

80 × 80	3.6	8.59	10.9	106	164	38.5
	5.0	11.7	14.9	139	217	49.8
	6.3	14.4	18.4	165	261	58.8
90 × 90	3.6	9.72	12.4	154	237	49.7
	5.0	13.3	16.9	202	315	64.9
	6.3	16.4	20.9	242	381	77.1
100 × 100	4.0	12.0	15.3	243	361	68.2
	5.0	14.8	18.9	283	439	81.9
	6.3	18.4	23.4	341	533	97.9
	8.0	22.9	29.1	408	646	116
	10.0	27.9	35.5	474	761	134
120 × 120	5.0	18.0	22.9	503	775	122
	6.3	22.3	28.5	610	949	147
	8.0	27.9	35.5	738	1159	176
	10.0	34.2	43.5	870	1381	206

For information (and certain limitations) in regard to design of members, reference should be made to 'Structural Steelwork Handbook for Standard Metric Sections: Structural Hollow Sections to BS 4848: Part 2' (CONSTRADO).

RECTANGULAR HOLLOW SECTIONS

DIMENSIONS AND PROPERTIES

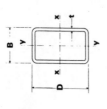

Designation		Mass per metre	Area of section	Moment of inertia		Torsional constants	
Size D × B	Thickness t			Axis x-x	Axis y-y	J	C
mm	mm	kg	cm²	cm⁴	cm⁴	cm⁴	cm³
50 × 30	2.6	3.03	3.86	12.4	5.45	12.1	5.90
	3.2	3.66	4.66	14.5	6.31	14.2	6.81
60 × 40	3.2	4.66	5.94	28.3	14.8	30.8	11.8
	4.0	5.72	7.28	33.6	17.3	36.6	13.7
80 × 40	3.2	5.67	7.22	58.1	19.1	46.1	16.1
	4.0	6.97	8.88	69.6	22.6	55.1	18.9
90 × 50	3.6	7.46	9.50	99.8	39.1	89.3	25.9
	5.0	10.1	12.9	130	50.0	116	32.9
100 × 50	3.2	7.18	9.14	117	39.1	93.3	26.4
	4.0	8.86	11.3	142	46.7	113	31.4
	5.0	10.9	13.9	170	55.1	135	37.0

Designation		Mass per metre	Area of section	Moment of inertia		Torsional constants	
Size D × B	Thickness t			Axis x-x	Axis y-y	J	C
mm	mm	kg	cm³	cm⁴	cm⁴	cm⁴	cm³
160 × 80	5.0	18.0	22.9	753	251	599	106
	6.3	22.3	28.5	917	302	729	127
	8.0	27.9	35.5	1113	361	882	151
	10.0	34.2	43.5	1318	419	1041	175
200 × 100	5.0	22.7	28.9	1509	509	1202	172
	6.3	28.3	36.0	1851	618	1473	208
	8.0	35.4	45.1	2269	747	1802	251
	10.0	43.6	55.5	2718	881	2154	296
	12.5	53.4	68.0	3218	1022	2541	342
	16.0	66.4	84.5	3808	1175	2988	393
250 × 150	6.3	38.2	48.6	4178	1886	4049	413
	8.0	48.0	61.1	5167	2317	5014	506
	10.0	59.3	75.5	6259	2784	6082	606
	12.5	73.0	93.0	7518	3310	7317	717
	16.0	91.5	117	9089	3943	8863	851

100 × 60	3.6	8.59	10.9	147	65.4	142	35.6
	5.0	11.7	14.9	192	84.7	187	45.9
	6.3	14.4	18.4	230	99.9	224	53.9
120 × 60	3.6	9.72	12.4	230	76.9	183	43.3
	5.0	13.3	16.9	304	99.9	242	56.0
	6.3	16.4	20.9	366	118	290	66.0
120 × 80	5.0	14.8	18.9	370	195	401	77.9
	6.3	18.4	23.4	447	234	486	93.0
	8.0	22.9	29.1	537	278	586	110
	10.0	27.9	35.5	628	320	688	126
150 × 100	5.0	18.7	23.9	747	396	806	127
	6.3	23.3	29.7	910	479	985	153
	8.0	29.1	37.1	1106	577	1202	184
	10.0	35.7	45.5	1312	678	1431	215

300 × 200	6.3	48.1	61.2	7880	4216	8468	681
	8.0	60.5	77.1	9798	5219	10549	840
	10.0	75.0	95.5	11940	6331	12890	1016
	12.5	92.6	118	14460	7619	15654	1217
	16.0	117	149	17700	9239	19227	1469
400 × 200	10.0	90.7	116	24140	8138	19236	1377
	12.5	112	143	29410	9820	23408	1657
	16.0	142	181	36300	11950	28835	2011
450 × 250	10.0	106	136	37180	14900	33247	1986
	12.5	132	168	45470	18100	40668	2407
	16.0	167	213	56420	22250	50478	2948

For information (and certain limitations) in regard to design of members, reference should be made to 'Structural Steelwork Handbook for Standard Metric Sections: Structural Hollow Sections to BS 4848: Part 2' (CONSTRADO).

REFERENCES

1. C. R. G. Hatton, Redesign stops connecting rod failures, *Engineering*, **183** (1959) 685
2. P.D. Swales and P. M. Braiden, A case history, *Engng Matls Design*, **17**, (1973) 30
3. P. Polak, *A Background to Engineering Design* (Macmillan, 1976)
4. T. von Karman and C. W. Zwang, Torsion with variable twist, *J. Aero. Sci.* **13** (1946) 10
5. R. J. Roark, *Formulas for Stress & Strain* (McGraw-Hill, 1976)
6. Helical Springs, *Engineering Design Guide 08* (Oxford University Press, 1974)
7. A. A. Griffith, The phenomena of rupture and flow in solids, *Phil. Trans. R. Soc. A*, **221**, 163 (1921) this reference p. 193 para. 5
8. *Engineering*, **11** (1871) 349 (see also other reports in the same volume)
9. J. K. Musuva and J. C. Radon, The effect of stress ratio and frequency on fatigue crack growth, *Fatigue of Engineering Materials & Structures*, **1** (1979) 457
10. M. B. Coyle and S. J. Watson, Fatigue strength of shafts with shrunk-on discs, *Proc. Inst. Mech, Engrs*, **178** (1963) 147
11. N. E. Frost, A relation between the critical alternating propagation stress and crack length [*sic*] for mild steel, *Proc. Inst. Mech. Engrs*, **173** (1959) 811
12. Motor Industry Research Association, The effect of heat cycling and ageing on the fatigue strength of fillet rolled components, *Report 1962/5* (M.I.R.A., Lindley, Nuneaton, Warwickshire, 1965)
13. P.G. Forrest, *Fatigue of Metals* (Pergamon, 1962)
14. R. B. Heywood, *Designing against Fatigue* (Chapman & Hall, 1962)
15. M. A. Miner, Cumulative damage in fatigue, *Trans. Am. Soc. Mech. Engrs*, **67** (1945) A 150
16. A. Palmgren, Die Dauerfestigkeit von Kugellagern, $VDI^{\#}Z$, **68** (1924) 339
17. H. O. Fuchs and R. I. Stephens, *Metal Fatigue in Engineering* (Wiley, 1980).
18. R. O. Ritchie, J. R. Rice and J. F. Knott, On the relationship between critical tensile stress and fracture toughness in mild steel, *J. Mech. Phys. Solids*, **21** (1973) 395

19. J. B. Burke (ed.), *Application of Fracture Mechanics to Design* (Plenum Press, 1979) 152ff.

20. P. L. Pratt (ed.), *Fracture 1969* (Chapman & Hall, 1969) p. 775

21. E. F. Church, *Steam Turbines* (McGraw-Hill, 1962)

22. R. E. Peterson, *Stress Concentration Factors* (Wiley, 1974)

23. I. M. Allison, Stress concentration in shouldered shafts, *Aeronaut. Q.* **12** 223;**13** 133 (1961–2)

24. R. E. Peterson, Model testing as applied to strength of materials *Trans. Am. Soc. Mech. Engrs* **55** (1933) APM 79 (fig. 7, p. 81)

25. R. Kuhnel, Achsbruche bei Eisenbahnfahrzeugen und ihre Ursachen, *Stahl Eisen*, **52** (1932) 965

26. R. E. Peterson and A. M. Wahl, Fatigue of shafts at fitted members, *Trans. Am. Soc. Mech. Engrs*, **57** (1935) A1

27. Various data items, Engineering Sciences Data Unit, 251 Regent St, London W1R 7AD

28. M. Hetenyi, Some applications of photoelasticity in turbine-generator design, *Trans. Am. Soc. Mech. Engrs*, **61** (1939) A 153–4

29. Data Item 68045, ESDU (see reference 27)

30. Data Item 68045, ESDU (see reference 27)

31. W. C. Orthwein, A new key and keyway design, *Trans. Am. Soc. Mech. Engrs*, **101** (1979) 383

32. H. Fessler, C. C. Rogers and P. Stanley, Stresses at keyway ends near shoulders, *J. Strain Analysis*, **4** (1969) 267

33. K. R. Rushton, Shouldered shaft stress concentration in torsion, *Aero. Q.*, **15** (1964) 95

34. A. G. M. Michell, The limit of economy of material in frame structures, *Phil. Mag.*, **8** (1904) 589

35. V. Leontovich, *Frames and Arches* (McGraw-Hill, 1959)

36. C. G. Anderson, Flexural stresses in curved beams of I and box section, *Proc. Inst. Mech. Engrs*, **163** (1950) 295

37. Neg'ator spring, Eastern Metals Research Inc., New York; UK licensee Tensator Ltd, Acton Lane, London NW10

38. F. A. Votta, Theory and design of long deflection constant force spring elements, *Trans. Am. Soc. Mech. Engrs*, **74** (1952) 439

39. W. Kloth, *Atlas der Spannungsfelder in Technischen Bauteilen* (Stahleisen, Dusseldorf, 1961)

40. Inquiry into the Basis of Design and Method of Erection of Steel Girder Bridges (Appendix 1 Part II, HMSO, London, 1973)

41. R. T. Gurney, *Fatigue of Welded Structures* (Cambridge University Press, 1968)

42. Data Item 71005, ESDU (see reference 27)

43. E. H. Gaylord and C. N. Gaylord, *Design of Steel Structures* 2nd ed. (McGraw-Hill, 1972)

44. P. B. Haigh, *Proc. Inst. Naval Arch.* **75** (1933)

45. J. B. Burke (ed.), *Application of Fracture Mechanics to Design* (Plenum Press, 1979) pp. 184–6

46. BS 5400 Steel, Concrete and Composite Bridges: Part 10, 1980, Code of Practice for Fatigue

47. J. B. Reber, Ultimate Strength Design of Tubular Joints, *Offshore Technology Conference* (Dallas, Texas 1972)

48. *Structural Steelwork Handbook* (British Construction Steelwork Association (London)

49. ASME Boiler and Pressure Vessel Code part VIII divisions 1 & 2 (American Society of Mechanical Engineers, New York)

50. BS 5500 1976 Unfired Fusion-welded Pressure Vessels

51. J. Adachi and M. Benicek, Buckling of torispherical shells under internal pressure, *Expl Mech.* **4** (1964), 217

52. G. E. Findlay, D. G. Moffatt and P. Stanley, *J. Strain Anal.* **3** (1968) 214

53. H. S. Tsien, Buckling of spherical shells under external pressure, *J. Aero, Sci.*, **9** (1943) 383

54. M. B. Bickell and C. Ruiz, *Pressure Vessel Design and Analysis* (Macmillan, 1967)

55. H. J. Bernhardt, Flange theory and the revised standard BS1 10 1962, *Proc. Inst. Mech. Engrs*, **178** pt. 1 (1963–64) 107

56. F. A. Leckie and R. K. Penny, *Welding Research Council Bulletin 90* (Welding Research Council, New York, 1963)

57. T. E. Pardue and I. Vigness, Properties of thin-walled curved tubes of short bend radius, *Trans. ASME*, **73** (1951) 77

58. R. T. Smith and H. Ford, Experiments on pipelines and pipe bends subjected to 3-dimensional loading, *J. Mech. Eng. Sci*, **9** (1967) 124

59. J. F. Whatham and J. J. Thompson, Bending and pressurising of pipe bends with flanged tangents, Nuclear Engng Des., **54**, (1979) 17

60. J. A. Haringx, Instability of bellows subjected to internal pressure, *Philips Res. Rep.*, **7** (1952) 189

61. D. E. Newland, Buckling of double bellows expansion joints under internal pressure, *J. Mech. Eng. Sci.* **6** (1964) 270

62. Department of Employment, *The Flixborough Disaster* (HMSO, 1974)

63. Civil Aircraft Accident Report, *Comet G–ALYP 10.1.54* and *Comet G–ALYY 8.4.54* (HMSO, 1955)

64. P. Polak, *Systematic Errors in Engineering Experiments* (Macmillan, 1978) p. 34

65. C. E. Phillips and R. B. Heywood. The size effect in fatigue of plain and notched steel specimens loaded under reversed direct stress, *Proc. Inst. Mech. Engrs.* **165** (1951) 113

66. K. J. Miller and K. P. Zachariah, Cumulative damage laws for fatigue crack initiation and stage I propagation, *J. Strain. Anal.*, **12** (1977) 262

67. R. W. Baldi, Experimental investigation of fatigue damage accumulation, *Aircr. Engng.*, **4** (1972) 14

68. J. B. Burke (ed.), *Applications of Fracture Mechanics to Design* (Plenum Press, 1979) p. 49

69. T. E. Stanton and R. G. C. Batson, On the characteristics of notched bar impact tests, *Proc. Inst. Civ. Engrs*, **211** (1920) 91/2 etc.

70. L. F. Kooistra, E. A. Lange and A. G. Pickett, Full-size pressure vessel testing and its application of design, *Trans. Am. Soc. Mech. Engrs*, **86** (1964) 419

71. A. J. Coker (ed.), *Automobile Engineer's Reference Book* (Newnes, 1959)

72. S. Timoshenko and D. H. Young, *Engineering Mechanics* (McGraw-Hill, 1959)

73. C. J. Smithells, *Metals Reference Book* (Butterworths, 1976)

74. V. K. Kinra and B. Q. Vu, Brittle fracture of plates in tension, virgin waves and boundary reflections, *Trans. Am. Soc. Mech. Engrs J. Appl. Mech.*, **47** (1980) 45

75. L. J. Hart-Smith, Buckling of thin cylindrical shells under uniform axial compression, *Int. J. Mech. Sci.* **12** (1970) 299

76. P. Stanley and T. D. Campbell, Very thin torispherical pressure vessel ends under internal pressure, *J. Strain Analysis* **16** (1981) 171

77. G. D. Galletly, Buckling and collapse of thin internally-pressurised dished ends, *Proc. Inst. Civ. Engrs* **67** pt 2 (1979) 607

INDEX